THE ESSENCE OF
ANALOG ELECTRONICS

THE ESSENCE OF ENGINEERING SERIES

Published titles
The Essence of Solid-State Electronics
The Essence of Electric Power Systems
The Essence of Measurement
The Essence of Engineering Thermodynamics
The Essence of Analog Electronics
The Essence of Power Electronics

Forthcoming titles
The Essence of Optoelectronics
The Essence of Microprocessor Engineering
The Essence of Communications
The Essence of Digital Design

THE ESSENCE OF

ANALOG ELECTRONICS

Colin Lunn

PEARSON
Prentice
Hall

Harlow, England • London • New York • Boston • San Francisco • Toronto • Sydney • Singapore • Hong Kong
Tokyo • Seoul • Taipei • New Delhi • Cape Town • Madrid • Mexico City • Amsterdam • Munich • Paris • Milan

07-529

Pearson Education Limited
Edinburgh Gate
Harlow
Essex CM20 2JE
England

and Associated Companies throughout the world

Visit us on the World Wide Web at:
http://www.pearsoned.co.uk

First published 1997 by
Prentice Hall Europe

© Prentice Hall Europe 1997

Typeset in 10/12pt Times
by MHL Typesetting Ltd, Coventry

Printed and bound by Antony Rowe
Ltd, Eastbourne

Library of Congress Cataloging-in-Publication Data

Lunn, Colin.
    The essence of analog electronics / Colin Lunn.
        p.   cm. — (The essence of engineering series)
    Includes bibliographical references and index.
    ISBN 0-13-360223-0
    1. Electronics.   I. Title.   II.   Series: Essence of engineering.
    TK7816.L86   1997
621.3815–dc20   95-25456
                                                        CIP

British Library Cataloguing in Publication Data

A catalogue record for this book is available from
the British Library

ISBN 0-13-360223-0

10   9   8
07   06   05   04

Contents

Preface

The essence of analog electronics is an introductory textbook, written for students in higher education who are studying electronics for the first time. Although aimed primarily at first-year undergraduate electronic engineering courses, it could easily be adopted for use in any technical courses such as BTEC HNC/HND where a knowledge of analog electronics is required. The book is specifically designed to cover material for a typical one-semester course.

The book covers the introductory electronics work necessary for students to progress to more advanced coverage of analog and digital units. Although confining itself mainly to analog devices and circuits, the text covers areas where there are distinct links with digital electronics, especially through switching circuit design and MOSFET inverting circuits

There are many comprehensive books already available in this area, but in the author's experience many students find their contents difficult to understand and the quantity of technical information overwhelming. This book is written specifically for students new to the subject and introduces each topic step by step Emphasis is given as much to understanding a concept as to being able to grasp the theoretical principles behind it.

In most topics, the theory is simplified by the use of realistic engineering approximations, which result in design rules which are easy to remember Design examples and case studies are given to illustrate the use of these rules in practice.

To enable students to monitor their own progress, each topic contains a set of self-assessment questions (SAQs). These may be found part way through a topic, to help students on their way, or at the end to establish the student's understanding of the work covered. The answers to all SAQs (which can be found at the end of the book) form part of the teaching process. In most cases the answers cover full worked solutions, helping and guiding the student through the process Extra information is occasionally given concerning other possible (or likely) answers and why they might, or might not, be acceptable.

By attempting the SAQs and checking the answers, the student's confidence will be built up sufficiently to attempt the tutorial questions at the end of each chapter. The answers to these questions are not given, but essential practice will be gained if students attempt them.

The book is written in an informal style which helps to give it a tutorial quality and hopefully makes the reader feel more involved in the text

The introductory nature of the book precludes any in-depth or advanced study of the topics covered and, for this reason, a guide to further reading is given at the end of each chapter for students wishing to extend their knowledge.

Although the mathematical content of the book is minimized where possible, it would be unrealistic to assume that a student could study engineering without using mathematics For this reason, it is assumed that students will have studied a foundation level course (or equivalent) in mathematics.

The structure and content of the book is strongly based on a teaching pattern of approximately one chapter per week. It has been found that certain topics may take students a little longer to grasp than has been planned for. To allow for this, the book contains ten chapters, giving some flexibility in a typical 12 week semester course.

The aim of Chapter 1 is to cover briefly the basic principles and vocabulary used in the rest of the book It is not the intention of this text to cover this work in great detail and it is recommended that students having no previous knowledge of these basics should refer to books on electrical circuits for more detailed explanations From Chapter 2 onwards, topics are covered which would be largely new to students.

Operational amplifiers are introduced early in the book and are covered in some detail over four chapters. Chapters 3, 4 and 6 cover the most popular operational amplifier circuits, while Chapter 5 focuses on some of the specifications and limitations of real operational amplifier chips. This early treatment of operational amplifiers not only emphasizes their importance in analog circuit design, but also gives students the tools to design and build circuits for specific applications early in their course. Operational amplifier circuit theory, being easy to approximate and rule-base, also helps to introduce some of the basic principles in circuit design without being too mathematical

Diodes, power supplies, other diode applications and zener voltage regulators are covered in Chapter 7. Integrated circuit regulators are also covered here to indicate their popularity over the zener versions.

Bipolar junction transistors, d c. and a.c. circuit designs are given in Chapters 8 and 9. Circuit approximations and simple rules are again used to demonstrate quick circuit design of voltage amplifiers. The theoretical treatment is based on the r_e model for a.c. circuits.

The popularity of MOSFETdevices is indicated by their coverage in Chapter 10 In this chapter there is a clear and important crossover into digital electronics, where MOSFET inverters, NMOS, PMOS and CMOS are explained. Owing to the steady decline in use of JFETdevices and the limited nature of a text of this size, coverage of JFETs has been omitted. Students requiring information on JFETs will, however, still find the text on MOSFET devices of use.

It is recommended that this book be used in conjunction with a laboratory course covering perhaps a major chapter topic per week. There can be no better way of learning and developing an intuition for electronics than by practical application

It is to be hoped that students will find this book interesting, easy to read and, most important, a launch pad for an extended interest in electronics

CHAPTER 1

Introduction

Aims and objectives

The aim of this introductory chapter is to give you some basic information concerning electronics Some of it you will know and some of it you will not have seen before. Hopefully there is a bit of both! If you find that you know all the things mentioned in this chapter, then by all means move on to the next chapter If, alternatively, you find much of the work new to you, then now is your chance to learn, before entering the world of electronics

After studying this chapter you will know about:

- *definitions of volt, ampere and power*
- *simple use of Kirchhoff's laws*
- *resistors*
- *capacitors*
- *inductors*
- *Ohm's law*
- *voltage and current dividers*
- *ideal and real voltage and current sources*

1.1 Voltage, current and power

We cannot venture very far into the world of electronics without some understanding of voltage, current and power. Let us start, therefore, with one or two definitions

1.1.1 Voltage

Voltage (V) is the amount of work done or the energy required (in joules) in moving a unit of positive charge (one coulomb) from a negative point (lower potential) to a more positive point (higher potential) and is measured in volts (V) Voltage is also called 'potential difference' (p.d.) and 'electromotive force' (e m f)

One joule of energy is required to move one coulomb of charge through a potential difference of one volt

$$1 \text{ volt} = 1 \text{ joule/coulomb} = 1 \text{ V}$$

1.1.2 Current

Current (I) is the amount of electric charge (coulombs) flowing past a specific point in a conductor over an interval of one second and is measured in amperes (A)

If one coulomb of charge flows past a specific point in a conductor over an interval of one second, then a current of one ampere flows

$$1 \text{ ampere} = 1 \text{ coulomb/second} = 1 \text{ A}$$

Note these extra points concerning current flow·

1   Although electric current is created by the flow of negatively charged electrons, we choose *by convention* to think of current flowing in the *opposite* direction to the flow of electrons (see Figure 1.1). This may seem odd at first, but the reason for this conventional current flow direction stems from the belief of early scientists that all current flow was the result of positive moving charge It was not until 1897, when Sir J. J Thomson discovered that the electron held negative charge, that the true current flow direction was realized For reasons of consistency, the original or conventional current direction is still used by electrical and electronic engineers.

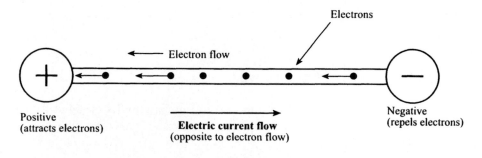

Figure 1.1

2.   Often in electronics the ampere (A) is too large a unit for convenience; it is far more common to use units of one thousand times smaller – milliamperes (mA) – or one million times smaller – microamperes (μA)

Now we have our basic definitions, we can look at how the terms 'voltage' and 'current' are used in practice.

1.1.3 Terminology of voltage

A voltage is a measure of the difference in potential across a component In other words, it is a measure of a potential relative to another potential You therefore need two points to measure a voltage Look at Figure 1 2

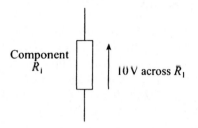

Component
R_1 10 V across R_1

Figure 1.2

There is, however, another way of describing voltage in a circuit If a voltage or potential difference is to be measured between a point on a circuit and the earth terminal, we can say 'the voltage at point A is X volts' (see Figure 1.3) This terminology is good shorthand if many voltages are being measured against one common reference point. A standard common reference point often used is called

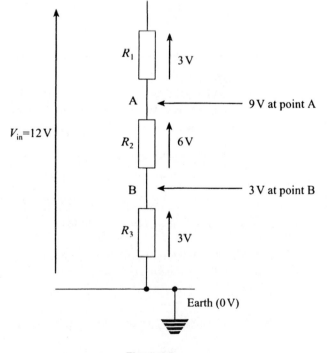

$V_{in}=12V$

R_1 3 V

A   9 V at point A

R_2 6 V

B   3 V at point B

R_3 3 V

Earth (0 V)

Figure 1.3

an earth. An earth point is always considered to be at zero potential Television engineers might measure voltages with respect to a common chassis voltage – which itself is sometimes riding at a much higher voltage than earth!

1.1.4 Terminology of current

As current is a flow of electrons, it can only be said to flow into, out of or through a component, as in Figures 1 4 and 1 5 For instance, different ways of describing current flow might be. 'a current of 10 A is flowing through the lamp'; 'the lamp is carrying a 10 A current', 'a 10 A current flows into the lamp', 'a 10 A current flows out of the lamp', 'we have 10 A flowing through the lamp'.

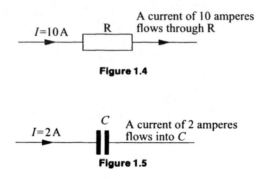

$I=10\,\text{A}$ R   A current of 10 amperes flows through R

Figure 1.4

$I=2\,\text{A}$ C   A current of 2 amperes flows into C

Figure 1.5

1.1.5 Power (P)

We have already described voltage as being the energy required to move charge, and current as the rate of flow of charge We can now go one step further and define a unit which measures the rate at which energy is used by an element This is called power and is measured in watts (W)

$$1 \text{ watt} = 1 \text{ joule/second} = 1 \text{ W}$$

We already know that 1 V = 1 joule/coulomb and that 1 A = 1 coulomb/second Therefore, if both units are multiplied together, we have

$$(1 \text{ V}) \times (1 \text{ A}) = (1 \text{ joule/coulomb}) \times (1 \text{ coulomb/second})$$
$$= 1 \text{ joule/second}$$
$$= 1 \text{ W}$$

Power is therefore simply related to voltage and current by the relationship

$$\text{power} = \text{voltage} \times \text{current}$$

i e

$$P = V \times I$$

1.2 Simple understanding of Kirchhoff's current and voltage laws

Kirchhoff's laws are usually given a thorough treatment in books on electric circuit analysis Although they are of great importance in circuit analysis, our intention here is to recognize the simple outcome of applying these laws, rather than to enter into the mathematical detail of their application to a circuit. There are two laws one concerning voltage and one concerning current

1.2.1 Kirchhoff's voltage law

Kirchhoff's voltage law is usually stated as 'The algebraic sum of the voltages around any closed circuit path is zero' More simply, it tells us that all components connected in parallel have the same voltage across them You can see that in Figure 1 6 we have a rather severe-looking circuit but, in terms of voltage, it simply contains lots of parallel components, each of which has the same voltage drop In other words, 'the voltage (or potential difference) across each component is equal'

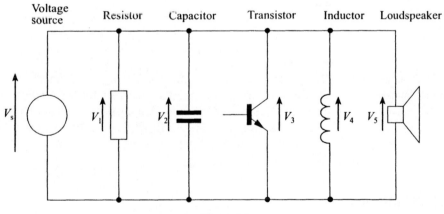

Figure 1.6

We can say, therefore,

$$V_s = V_1 = V_2 = V_3 = V_4 = V_5$$

1.2.2 Kirchhoff's current law

Kirchhoff's current law is usually stated as 'The algebraic sum of currents flowing into or out of a junction is zero' This can be restated as 'The sum of currents flowing into a junction is equal to the sum of currents flowing out a junction'

All we need to understand by this law is that all components connected in series have the same current flowing through them. This is illustrated in the circuit of Figure 1.7, which, although not exactly simple, does allow us to make a simple observation concerning the current flow in the individual components. We can say that

$$I_s = I_r = I_d = I_i = I_{spk} = I_u = I_f$$

meaning that each series component carries the same current.

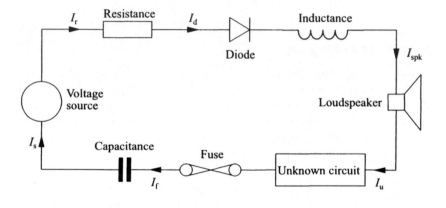

Figure 1.7

You may wonder, 'Surely there's more to Kirchhoff's laws than this?' You would be correct, but the same concept applies, it is just that further studies of circuit analysis (not possible in this text) would demonstrate more complex applications of both the laws. The simple outcome of each law as shown here will help tremendously in our understanding of circuit operation.

1.3 Resistors

Components which resist or reduce the flow of current in a circuit are called resistances. A practical resistance is called a 'resistor' and, in a circuit, exhibits its properties of resistance better than any other property it may have. The unit of resistance is the ohm (Ω).

1.3.1 Resistors in series

When we connect resistors in series, the total resistance is simply the sum of the resistances of the individual resistors. Look at Figure 1.8 to get the idea.

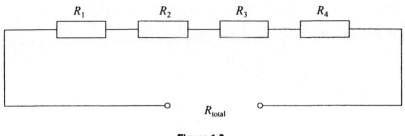

Figure 1.8

The total resistance is

$$R_{total} = R_1 + R_2 + R_3 + R_4$$

1.3.2 Resistors in parallel

To find the total resistance of resistors connected together in parallel, we must normally take the sum of the reciprocals of the individual resistors, which gives us the reciprocal of the total. Of course, we then take the reciprocal of this value to determine the final total resistance R_{total} If you refer to Figure 1 9, you will see that

$$1/R_{total} = 1/R_1 + 1/R_2 + 1/R_3 + 1/R_4$$

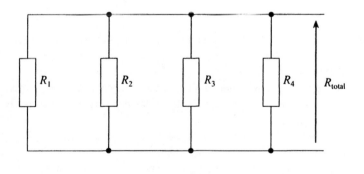

Figure 1.9

EXAMPLE 1.1
If each resistor had the value 68 Ω, find R_{total}.

SOLUTION
R_{total} would be found from

$$1/R_{total} = 1/68 + 1/68 + 1/68 + 1/68$$
$$= 4(1/68)$$
$$= 4/68$$

Therefore

$$R_{total} = 68/4 = 17\Omega$$

A special point about parallel resistances (which is not so obvious at first sight) is that the total parallel resistance R_{total} is *always* dominated by the smallest resistance value in the circuit. To illustrate this, let us work out the R_{total} value of this example if one of the $68\,\Omega$ resistors were replaced by a $1\,\Omega$ resistor We would have

$$1/R_{total} = 1/1 + 1/68 + 1/68 + 1/68$$
$$= (68/68) + (3/68)$$
$$= 71/68$$

Therefore

$$R_{total} = 68/71$$
$$\approx 0.96\,\Omega$$

Note that the value of R_{total} has dropped to less than $1\,\Omega$

Quick check rule
To help you to perform quick checks on resistance circuits, it is worth remembering that the total resistance of a parallel resistor network is always dominated by, and is less than, the smallest resistor.

Product-over-sum rule for two resistances
Engineers would normally use the product-over-sum rule for two parallel resistors You may have heard about this already It gives us a short-cut method for finding R_{total}

Consider two resistors R_1 and R_2 in parallel. We can say

$$1/R_{total} = 1/R_1 + 1/R_2$$

Using a common denominator of $(R_1 \times R_2)$ for the right-hand side of this equation, we have

$$1/R_{total} = [R_2/(R_1 \times R_2)] + [R_1/(R_1 \times R_2)]$$

Therefore

$$1/R_{total} = (R_2 + R_1)/(R_1 \times R_2)$$

We can now invert both sides of the equation to yield R_{total}

$$R_{total} = R_1 \times R_2/(R_1 + R_2)$$

In other words

$$R_{total} = \text{product of resistors/sum of resistors}$$

However, if we have three resistances, the mathematics gives an awkward result

$$R_{total} = (R_1 \times R_2 \times R_3)/(R_2 \times R_3 + R_1 \times R_3 + R_1 \times R_2)$$

Although this expression could be described as being in the form of product over sum, it is clearly not an easy one to remember: in fact, it is rather more involved than the standard reciprocal relationship It is therefore not a good idea to use the product-over-sum method for more than two resistances There is nothing wrong, however, with using this method more than once – for each parallel pair of resistances in a large parallel resistance network. The choice is yours!

Using shorthand notation for parallel resistances
There is an accepted shorthand notation for indicating two or more resistors connected in parallel Instead of writing, say,

resistor R_1 is connected in parallel with resistor R_2

we would write

$$R_1 \parallel R_2$$

Note the use of two vertical parallel lines This tells the reader that the two resistors are connected in parallel. This is very useful shorthand when trying to determine the total combined resistance of a large network of resistances. For instance, in Figure 1 9 we could write

$$R_{total} = (R_1 \parallel R_2 \parallel R_3 \parallel R_4)$$

Uses of resistance in circuits
Resistances are used in circuits to control or limit the amount of current flow in a wire. We could, for instance, increase the size of the resistance in Figure 1 7 to reduce the maximum size of current flow in the circuit at any time

Another major use of a circuit resistance is to convert a flow of current into a voltage We will see shortly how Ohm's law describes the relationship between current flow through, and voltage across, a simple resistance. The resistor is the cheapest and simplest form of current-to-voltage converter available

Because of the resistor's ability to generate voltage from current flow, a voltage dropper is often used to reduce the size of a voltage from one circuit to another

An important point about circuit resistance is that, very often, resistances appear in circuits without our having added a resistance component. Every piece of copper wire, for instance, possesses a small but unwanted resistance value. A component or circuit element may often possess resistance. We would rather it did not, but we need to be aware of its existence.

Self-assessment questions

SAQ1.1 State what is wrong with the following statements:
1   A voltage reading was taken at the resistor R   .
2   . . . and the current at the bulb was found to be smaller than expected!
3. There was a voltage drop of 5.5 V through the resistance.

SAQ1.2 True or false?
1   A current flow is a measure of the rate of flow of charge (dQ/dt)
2. Conventional current flow is in the same direction as electron flow
3. Current flow is always from a negative node to a positive node.
4. The total resistance of a parallel resistor network is always larger than the largest individual resistance.
5. The sum of the currents leaving a point is equal to the sum of the currents arriving at the same point.
6. The above statement uses Kirchhoff's voltage law.

SAQ1.3 Three resistances, $10\,k\Omega$, $24\,k\Omega$ and $56\,k\Omega$, are connected first in series and then in parallel. What is the total resistance value in each case?

SAQ1.4 Prove that for two resistances in parallel the total resistance R_{total} is given by

$$R_{total} = (R_1 \times R_2)/(R_1 + R_2)$$

1.4 Capacitors

Components which have the ability to store electric charge (Q) in a circuit are called 'capacitances'. The larger the capacitance of a component, the more the amount (or capacity) of electric charge it can store

A practical capacitance is called a 'capacitor' and, in a circuit, exhibits its properties of capacitance better than any other property it may have The unit of capacitance is the farad (F). Finding total values for capacitance in series and parallel circuits is the exact opposite to the technique for finding total resistance values

1.4.1 Capacitors in parallel

When we connect capacitors in parallel, the total capacitance is simply the sum of the individual capacitors. Looking at Figure 1 10, we say that the total capacitance is

$$C_{\text{total}} = C_1 + C_2 + C_3 + C_4$$

Note the difference from the parallel resistor case

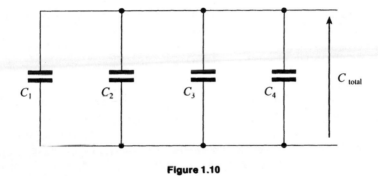

Figure 1.10

1.4.2 Capacitors in series

To find the total capacitance of capacitors connected together in series, we must take the sum of the reciprocals of the individual capacitors, which gives us the reciprocal of the total. To determine the final total capacitance, C_{total}, we then take the reciprocal of this value. If you refer to Figure 1.11, we can say that

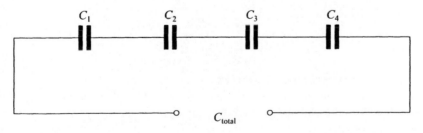

Figure 1.11

$$1/C_{total} = 1/C_1 + 1/C_2 + 1/C_3 + 1/C_4$$

Note that if we have only two capacitors, as in the case of parallel resistors, there is a short-cut method. In this case we can use the product-over-sum method

$$C_{total} = C_1 \times C_2/(C_1 + C_2)$$

1.4.3 More things about capacitors

Another piece of information we need to know about is the relationship between charge (Q) and capacitance (C) Capacitors are storage components for energy and charge. A large capacitor can store more charge than a small capacitor The relationship linking this idea is

$$Q = C \times V$$

where Q is the stored charge in coulombs, C is the capacity in farads and V is the voltage across the capacitor. You can also see from this that, if the voltage across the capacitor is increased, the amount of stored charge increases correspondingly

Capacitor current

It would be incorrect to talk about the current flowing through a capacitor, because the component has an insulating layer between its plates. We can, however, refer to capacitor current. There is a sneaky difference here in that charge can flow into one plate and a corresponding charge can leave the other plate, giving the appearance of a connected current flow *through* the device, although all that has really happened is that some charge has moved around an external circuit. In fact, the very definition of current is the rate of flow of charge. The mathematical way of indicating small changes in quantities is to write d before the amount. Therefore, a small change in charge Q would be written as dQ. Similarly, a small change in time is written as dt. The change in charge occurring over a time interval is written as dQ/dt and indicates a rate of change of charge in time t. We say, 'dQ by dt'. This type of representation of changes in quantities forms the basis of a mathematical study called 'differential calculus'. We can therefore write

$$I = dQ/dt$$

Finding the size of the capacitor current

The amount of current flowing at any time around a capacitor circuit is dependent on how quickly the voltage across the capacitor is changing In other words, because I = dQ/dt, by definition, and $Q = C \times V$, if the capacitance remains constant, we can say

$$I = C \, dV/dt$$

This tells us that, provided the capacity C is constant, we can control the amount of current by changing the voltage across the capacitor. A slow change of V will cause a small I to flow, whereas a rapid change of V will increase I.

Uses of capacitors in circuits

Capacitors are used extensively in circuits, in a number of different ways. The three most common applications listed below illustrate their flexibility. All capacitor applications, however, utilize the capacitor's ability to store electric charge

1   *Timing circuits.* Many circuits use capacitors because of their ability to charge or discharge slowly over a time interval through an external resistance.
2.   *Voltage blocking.* Because the charge on a capacitor (and therefore the voltage across a capacitor) cannot be *instantly* increased or reduced, capacitors are readily used to transfer rapidly changing voltages from one circuit to another and block slowly changing voltages from one circuit to another This leads to their use in frequency dependent circuits such as graphic equalizers and loudspeaker crossover units.
3   *Tuned circuits.* Combinations of capacitors and inductors (see Section 1.5) are often used to produce tunable frequency selective circuits as used in television and radio equipment

1.5 Inductors

If a length of wire is formed into a coil and its ends are connected to a voltmeter, it is found that when a magnet is moved through the coil, a voltage is induced across its ends In fact, the amount of induced voltage is dependent on how fast the magnet is moved through the coil If the ends of the coil are connected together to form a closed path, a current will flow around the coil. A magnet has a magnetic field surrounding it and, by moving the magnet through the coil, the magnetic field is moving with it This property of the wire coil to *induce* a voltage and a current flow within itself, from the effect of a moving magnet (or changing magnetic field), is called 'self-inductance' and explains why the coil is called an inductor. Very often, self-inductance is simply called 'inductance'.

The discovery of this link between electric and magnetic fields was made by Faraday in 1831. Later work by Lenz in 1835 led to a rule which described the *direction* of the induced current as 'always in such a direction as to oppose the change causing it'.

A practical inductance is called an 'inductor' and, in a circuit, exhibits its properties of inductance better than any other property it may have. The unit of inductance is the henry (H). As 1 H is rather a large value, millihenries (mH) are more commonly used. The symbol for inductance (or self-inductance) is L (the symbols I and S already being used for other quantities). We might say, for instance, that an inductor may have an inductance L of 100 mH.

1.5.1 Inductors in series

When we connect inductors in series, the total inductance is simply the sum of the individual inductors In Figure 1.12 we would find the total inductance is

$$L_{total} = L_1 + L_2 + L_3 + L_4$$

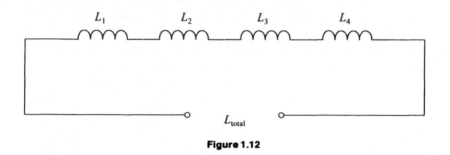

Figure 1.12

1.5.2 Inductors in parallel

To find the total inductance of inductors connected together in parallel, we must take the sum of the reciprocals of the individual inductors, which gives us the reciprocal of the total inductance. To determine the final total inductance, L_{total}, we then take the reciprocal of this value. If you refer to Figure 1.13, we can say that

$$1/L_{total} = 1/L_1 + 1/L_2 + 1/L_3 + 1/L_4$$

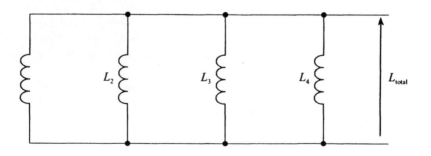

Figure 1.13

1.5.3 More things about inductors

Inductors are storage components for energy and magnetic field. When a current flows through an inductor, the voltage drop across the component is zero and a magnetic field is created which is a store of energy. When the current is cut off, the stored energy produces a voltage across the component.

Finding the size of the voltage appearing across the inductor
This voltage appears only while the stored energy is changing, i.e only when we are changing the current flow through the device. Therefore, the size of voltage V generated at any time across the inductor is dependent on how quickly the current through the inductor is changing If the rate of change of current is dI/dt (to use calculus representation), we can say

$$V = L \, dI/dt$$

This tells us that, provided the inductance L is constant, we can control the size of induced voltage by changing the current flow through the inductor A slow change of I will cause a small V to be created, whereas a quick change of I will increase V.

Uses of inductors in circuits
As with capacitors, the list of circuits utilizing inductors is an extensive one It should be mentioned here, however, that inductors are relatively expensive, sensitive, fragile and large components, and their use is often avoided. An inductor will be used in a circuit because of its ability to oppose changes in voltage across its ends or changes in current flow through it.

Frequency suppressors
As the inductor opposes rapid changes in voltage, it is often used to suppress (reduce) the amount of high frequency interference in a circuit

Tuned circuits
As with the capacitor, the inductor has frequency-dependent properties and is often found in broadcast transmitter and receiver circuits.

Frequency-selective filters
The frequency-dependent properties of the inductor lead to its wide use in circuits (usually in conjunction with capacitors) which filter, or separate out, different frequencies in a system

1.6 Impedances

Impedance (Z) is a generalized term for the property of a component containing any combination of the properties of resistance R, capacitance C and inductance L. Impedance, like resistance, is measured in ohms A component having any capacitance or inductance will have properties which are frequency-dependent. Very often in electronics we are not absolutely sure of the size of (or even the presence of) each of these properties and therefore we use the term 'impedance' as a generalization For instance, a coil of wire would normally be considered to be an inductance L, but the wire used to wind the coil itself will have a certain resistance value R (often quoted in ohms per metre). We are left with the small problem of whether to consider the coil as a resistance R or as an inductance L The answer is to consider it as a combination of both, i.e. an impedance Z

1.7 Ohm's law

If we take a constant resistance R and drive a current I through it, a voltage V is created across the resistance. If we then increase the current flow, we obtain a corresponding increase in voltage. This relationship between voltage and current is a linear one and is represented by the most fundamental and famous law in electronics, called 'Ohm's law', without which we would make little progress

$$I = V/R$$

Remembering it in words might help, 'I equals V over R'.

   Take a look at Figure 1 14 to get a picture in your mind about what Ohm's law is saying. Notice the polarity of the voltage. When current flows into a resistance, *that* end becomes the positive polarity We can (and often do) rearrange the Ohm's law equation into its voltage form

$$V = I \times R$$

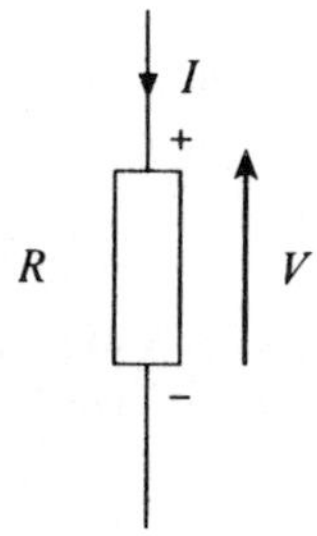

Figure 1.14

For this reason, a voltage difference (or voltage drop) is often referred to as an 'IR drop'.

   Try a quick example

EXAMPLE 1.2

If a resistance $R = 50 \ \Omega$ and a current $I = 1 \ \text{mA}$ flows through it, what voltage (potential difference) is developed across R?

SOLUTION

$$I = V/R$$

Therefore

$$V = I \times R$$

Therefore

$$V = (1 \times 10^{-3}) \times 50 = 50 \ \text{mV}$$

Self-assessment question

SAQ1.5 In Figure 1 7 which end of the resistance is positive?

You can see that the Ohm's law equation is simple, but let us see a mental arithmetic short cut. Quite often circuit values might be given as, say, $R = 1 \ \text{k}\Omega$ and $I = 1 \ \text{mA}$ This is the same as saying $R = 1 \times 10^3 \ \Omega$ and $I = 1 \times 10^{-3}$ A Since we know that the voltage $V = IR$, we would have

$$V = (1 \times 10^{-3}) \times (1 \times 10^{3}) = 1 \ \text{V}$$

Notice that the 10^{-3} cancels with the 10^3. Provided that we can recognize that the product of a kilohm and a milliamp is a volt, we can sometimes make the calculation a little easier for ourselves, as in Example 1.3.

EXAMPLE 1.3

If $R = 20 \ \text{k}\Omega$ and $I = 3 \ \text{mA}$, what is V?

SOLUTION

You could calculate this using $R = 20 \times 10^3 \ \Omega$ and $I = 3 \times 10^{-3}$ A, or you could recognize that you are taking the product of 10^3 and 10^{-3}, which cancel out, leaving

$$V = 20 \times 3 = 60 \ \text{V}$$

In the same way, you can mentally calculate without difficulty the value of I directly in mA.

EXAMPLE 1.4

Looking at Figure 1.14, if $V = 5 8$ V and $R = 2$ kΩ, what is I?

SOLUTION

We have volts/kΩ, which gives the answer directly in mA Therefore

$$I = 5.8 \text{ V} / 2 \text{ k}\Omega = 2.9 \text{ mA}$$

Self-assessment questions

SAQ1.6 What units do we get from volts/mA?

SAQ1.7 Using Figure 1 15, if R_L is chosen to be 50 kΩ.

1. What is the current in the 10 kΩ resistor?
2   What is the current in R_L?
3   What is the voltage across the 10 kΩ resistor?
4. What is the voltage across R_L?

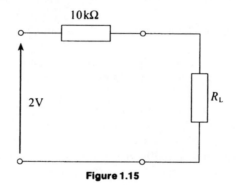

Figure 1.15

If you worked out the values for SAQ1 7, you no doubt used Ohm's law at least twice. It would be very useful if we could work out the voltage across R_L from only one equation, therefore necessitating only one calculation Fortunately, there *is* an equation to do this; it is called the 'voltage divider equation' It can be used whenever we see the voltage divider circuit

1.8 The voltage divider circuit and equation

The voltage divider is a quick way to find the voltage across one resistor when two or more resistors are connected in series. There are many analog circuits that make

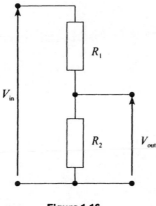

Figure 1.16

use of the voltage divider, shown in Figure 1.16 The expression for the output is derived very simply from Ohm's law It is

$$V_{out} = V_{in} \times R_2/(R_1 + R_2)$$

Notice that the output is merely a fraction of the input. The fraction is formed by the resistance ratio $R_2/(R_1 + R_2)$ This means that we can obtain any fraction of the input between 0 and 1 by choosing the appropriate resistor values Before you are tempted to memorize this equation, note that the resistance on the numerator R_2 is the resistance across which the output voltage is generated It is not so much the equation that is useful as the circuit itself. If a voltage is needed in a circuit anywhere, more often than not it is created from a voltage divider

For example, take a simple 12 V car battery If we needed to create any voltage in the range 0 V–12 V, we could use a voltage divider across the battery. Figure 1 17 shows how we can generate a 2 V output from a 12 V input. Note how the output is calculated

$$V_{out} = V_{in} \times R_2/(R_1 + R_2)$$

$$= 12 \times 4 \text{ k}\Omega/(20 \text{ k}\Omega + 4 \text{ k}\Omega)$$

$$= 48/24$$

$$= 2 \text{ V}$$

Be careful with the units Here kilohms are used for all R values, which cancel out. If you are not confident, use ohms throughout. Do not mix kilohms and ohms in the same calculation·

$$V_{out} = 12 \times 4000/(20\,000 + 4000) = 2 \text{ V}$$

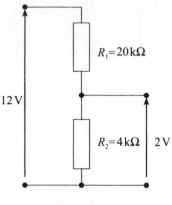

Figure 1 17

This of course means that the remaining 10 V is dropped across R_1 You can see quite easily that the voltage divider circuit just splits the V_{in} in the ratio R_1 R_2, which is the same as the ratio V_{R1} V_{R2} In the case shown here, 20 000   4000 gives 10 V $\cdot$ 2 V You can always use this as a check after a calculation

For the more general case of voltage dividers using more than two resistors, the output voltage expression should have said

$$V_{out} = V_{in} \times R_2/(\text{total series resistance})$$

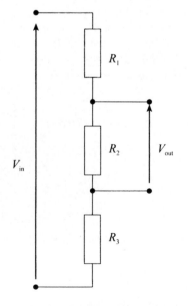

Figure 1 18

or to be more general still

$$V_{out} = V_{in} \times \frac{\text{the resistance across which the output voltage is read}}{\text{the total series resistance across } V_{in}}$$

Therefore, for a circuit like the one shown in Figure 1.18, we would use

$$V_{out} = V_{in} \times R_2/(R_1 + R_2 + R_3)$$

You can try some questions a little later on this We can now look at the second of the divider circuits in common use· the current divider.

1.9 The current divider circuit and equation

Used just as extensively as the voltage divider, the current divider is a useful circuit with an equation in a similar form to that of the voltage divider. If you want a known fraction of a current, or if you want to determine the current flow in a parallel resistance circuit, then the current divider is for you! The circuit is shown in Figure 1.19.

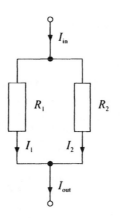

Figure 1.19

To find the currents I_1 and I_2, we can apply Ohm's law First, we simplify the calculation by combining the two parallel resistors R_1 and R_2 to give

$$R_{total} = (R_1 \times R_2)/(R_1 + R_2)$$

The current I_{in} flows into R_{total} Therefore the voltage V across R_{total} is given by

$$V = I_{in} \times (R_1 \times R_2)/(R_1 + R_2)$$

As R_1 and R_2 are in parallel, they have the same voltage difference V across each of them. Therefore

$$I_1 = V/R_1$$

Substituting the value for V into this equation, we have

$$I_1 = [I_{in} \times (R_1 \times R_2)/(R_1 + R_2)]/R_1$$

We can cancel R_1 from this to give

$$I_1 = I_{in} \times R_2/(R_1 + R_2)$$

Similarly, we find

$$I_2 = I_{in} \times R_1/(R_1 + R_2)$$

A point to note is that to find the current in R_1 the equation uses the value of R_2 and to find the current in R_2 the equation uses the value of R_1

Now try a few questions to consolidate your grasp of these ideas.

Self-assessment questions

SAQ1.8 If $R_1 = 10\,k\Omega$, $R_2 = 25\,k\Omega$ and $I_{in} = 1.5\,mA$ in Figure 1 19

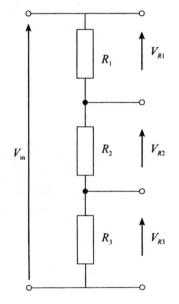

Figure 1.20

1   What is the voltage across R_1?
2. What is the voltage across R_2?
3. What are the values of I_1 and I_2?

SAQ1.9 With reference to Figure 1.16, find V_{out} for.

1. $V_{in} = 15$ V, $R_1 = 24$ kΩ and $R_2 = 12$ kΩ;
2   $V_{in} = 30$ V, $R_1 = 42$ kΩ and $R_2 = 3$ kΩ,
3   $V_{in} = 7$ V, $R_1 = 100$ kΩ and $R_2 = 250$ kΩ,
4   $V_{in} = 18$ V, $R_1 = 72$ kΩ and $R_2 = 9$ kΩ.

SAQ1.10 Look at the circuit in Figure 1 20 and answer the following questions.

1   Do R_1, R_2 and R_3 carry the same current?
2   If $V_{in} = 12$ V, $R_1 = 3$ kΩ, $R_2 = 18$ kΩ and $R_3 = 15$ kΩ, is $V_{R3} = 5$ V?
3   Does $V_{in} - (V_{R2} + V_{R1}) = V_{R3}$?

1.10 Voltage and current sources

We use sources of voltage and current all the time in electronics, so we will now look at what exactly is meant by a voltage source and a current source The important part of this work is to understand that we have a concept (a theoretical understanding) of an *ideal* source of voltage (or current), but in practice we only see and use *real* sources The only difference between a real and an ideal source is the size of the internal resistance of the source. Let us look at voltage first

1.10.1 Voltage sources

An ideal voltage source is one that provides a constant voltage across any circuit connected to it, as shown in Figure 1 21, where the circuit connected to the voltage source V_S is a simple resistance R_L In a circuit like this, it is generally accepted terminology to call a flow of current into load R_L 'a load current', i e the resistance is said to be 'loading' the voltage source and R_L is called a load resistance.

A real voltage source, however, will give us an output voltage which reduces as we reduce the size of R_L (the reason for this is explained shortly) This is shown in Figure 1.22.

The problem with a real voltage source is caused by the action of the internal resistance R_{int}, as shown in Figure 1.22 When current is drawn from the source, a voltage (IR) drop appears across R_{int}, leaving a reduced voltage available to drive the load R_L You might surmise from this that the ideal source has no internal resistance This is true, but to allow us to make a comparison between an ideal and a real voltage source, it is easier to assume that an ideal source *does* have an internal resistance, but its value is zero. Therefore, for an ideal voltage source

$$R_{int} = 0\,\Omega$$

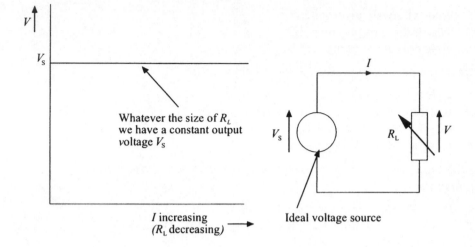

Whatever the size of R_L we have a constant output voltage V_s

I increasing
(R_L decreasing)

Ideal voltage source

Figure 1.21

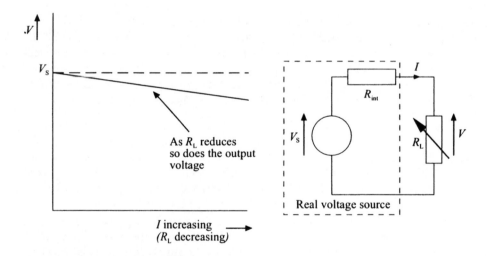

As R_L reduces so does the output voltage

I increasing
(R_L decreasing)

Real voltage source

Figure 1.22

whereas for a real voltage source

$$R_{int} > 0\,\Omega$$

These ideas are best illustrated by a real-life example. A commonly available voltage source in real life (and therefore a real voltage source) is a battery. Consider the following application of a real voltage source

CASE STUDY

Imagine that you have available a voltage source This is quite easy because you will have bought (at one time or another) a battery for your radio or lamp or such like Let us say that you have bought a 6 V battery for a portable lamp You fit the battery into the lamp and switch it on You also have a portable voltmeter which you use to measure the voltage across the two battery terminals (we will call this the 'battery terminal voltage'). What voltage do you read? 6 V of course! Okay, so you leave the lamp on for a while Then you read the battery voltage again. What has happened? Well, you find this time that the reading has dropped to, say, 4 V (The longer you leave the lamp on, the lower the terminal voltage becomes.) The odd thing about this voltage is that if, while reading the voltage you turn the lamp off, the voltage reading returns to 6 V! Figure 1 23 shows this example

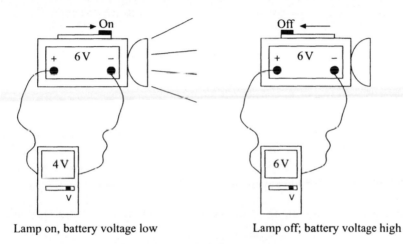

| Lamp on, battery voltage low | Lamp off; battery voltage high |

Figure 1.23

What has happened to the battery terminal voltage? The explanation is that the battery starts as an almost ideal voltage source with virtually zero internal resistance As the battery energy is used up, its internal resistance rises This causes the load current to the lamp to develop an internal voltage drop across the internal resistance. The voltage that is left (6 V − (internal voltage drop)) then appears at the battery terminals (See the circuit for this in Figure 1.24) You can easily see, therefore, that if the lamp is turned off, load current stops flowing, and the battery internal resistance carries zero current and therefore drops zero voltage Hence, when not loaded the battery terminals return to 6 V.

What happens when the battery goes flat? This would happen (as we all know to our cost) if we left the lamp switched on for a long period of time There is a point at which the battery internal resistance rises to such a large value that the battery terminal voltage becomes almost zero and the lamp fails to light

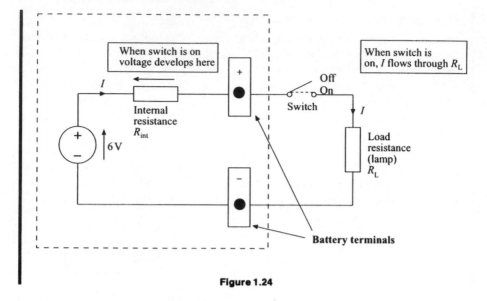

Figure 1.24

Self-assessment questions

SAQ1.11 If the lamp in the example quoted in the case study is switched on until the 6 V battery terminal voltage is 3.5 V, where has the lost 2 5 V gone? Could you read the lost 2.5 V directly by applying the voltmeter to a particular point in the circuit? (Hint: remember this is a battery.)

SAQ1.12 Again in the same example, assuming that the 6 V lamp requires a 0.5 A current at full power:

1   At the voltage and current given, what is the power of the lamp?
2   If the lamp has a 12 Ω resistance and the terminal voltage has dropped to 4 V after two hours' use, what is the internal battery resistance at this time?
3   What is the approximate internal resistance of a new battery?
4   What is the internal resistance of an ideal voltage source?

SAQ1.13 A piezo-crystal microphone is a very poor voltage source What, therefore, do you think happens to the terminal voltage signal when it is connected to a load resistance which draws a load current from the microphone?

1.10.2 Current sources

A good current source (an *I* source) is one that provides a constant current into any circuit (or load) connected to it.

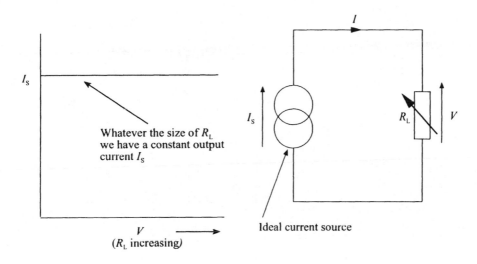

Whatever the size of R_L
we have a constant output
current I_s

V
(R_L increasing)

Ideal current source

Figure 1.25

In the section on voltage sources, an easily obtainable voltage source was introduced Current sources, however, are not so easily found, although they can be generated quite easily. They do what their name suggests provide a current They are produced to provide a constant current, irrespective of the circuits that may be connected to them (Compare this with voltage sources, which are used to provide a constant voltage across any circuit connected to them)

An ideal current source is one that provides a constant current through any load, as shown in Figure 1 25

A real current source, however, will give us an output current I which reduces as we increase the size of R_L. This is shown in Figure 1.26.

The problem with a real current source is caused by the action of the internal resistance R_{int}, which behaves as if an internal resistance had been connected across the terminals of the current source This is shown in Figure 1.26. The current I_S from the source splits into two smaller currents, I and I_{int}, leaving a reduced current ($I_S - I_{int}$) available to drive the load R_L As in the case of the ideal voltage source, you might surmise that the ideal current source has zero internal resistance This is wrong. We have to be a little careful here when considering what must be happening to the current flow in the ideal case Ideally, none of the current is lost inside the source Therefore, for this to be true we must have an infinitely large internal resistance To allow us to make a comparison between an ideal and a real current source, it is easier to assume that an ideal current source does have an internal resistance and that its value is infinitely high

Therefore, for an ideal current source

$$R_{int} = \infty\, \Omega$$

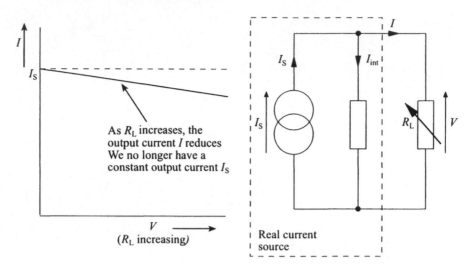

As R_L increases, the
output current I reduces
We no longer have a
constant output current I_S

V
(R_L increasing)

Real current
source

Figure 1.26

whereas for a real current source

$$R_{int} < \infty \, \Omega$$

Self-assessment questions

SAQ1.14 A 100 mA current source of internal resistance 10 kΩ is used to drive a 150 Ω load. What is the load current? If you wanted to increase the load current by changing the value of the 150 Ω load resistance, would you choose to increase it or decrease it?

SAQ1.15 What is the ideal internal resistance of a current source?

SAQ1.16 Which is a better current source: one with $R_{int} = 110$ kΩ or one with $R_{int} = 1$ kΩ?

You should now have a better idea about V and I sources. The reason for considering real current and voltage sources is to help us understand their limitations and to enable us to predict more accurately their effect on circuits using them.

This leads us nicely to the topics in the next chapter, where we consider what happens when you connect a real source to a variety of different types of load circuit.

Checklist

Do you know.
- the difference between electron flow and conventional current flow?
- the terminology of current and voltage measurements on a circuit?
- how to calculate parallel resistance values for two or more resistances?
- when you can use the product-over-sum rule?
- how to calculate series or parallel inductance values for two or more inductances?
- how to calculate series or parallel capacitance values for two or more capacitances?
- the relationship between current and voltage for a capacitor?
- the relationship between current and voltage for an inductor?
- the Ohm's law equation?
- how to draw a voltage divider circuit?
- how to draw a current divider circuit?
- the voltage divider equation?
- the current divider equation?
- the difference between a real and ideal voltage source?
- the difference between a real and ideal current source?
- the ideal internal resistances of voltage and current sources?

Tutorial questions

1   Correct the following statements·
- Resistor R has 4 mA across it
- The amperage through R_1 was 10 mA
- 240 V flows through the electric fire
2   Are the following statements true or false?
- A voltage can be measured at a single point in a circuit, without reference to any other point
- Electron flow and conventional current flow are in the same direction
- If current flows into a resistance from left to right, the right-hand end is the more negative end
- The total resistance of a series resistor network is always larger than any one individual resistance
- All components connected in parallel carry the same size current.
- All components connected in series carry the same size current
3   Three resistances of 1.5 kΩ, 3 9 kΩ and 1.5 kΩ are connected first in series and then in parallel What is the total resistance value in each case?
4   If the total series connection of resistances described in question 3 is connected across a 25 V supply, what is the power absorbed by each resistance?

5   By using the product-over-sum rule twice, find the total parallel resistance of the three resistances 110 kΩ, 39 kΩ and 68 kΩ.

6   By using only the product-over-sum rule three times, find the total combined parallel resistance of the four resistances 10 kΩ, 39 kΩ, 24 kΩ and 48 kΩ.

7   Repeat the same thing as asked for in question 6 for 4.8 Ω, 6.8 Ω, 9.2 Ω and 3.3 Ω

8   Two capacitors of 0 1 μF and 0.125 μF are connected first in series and then in parallel. What is the total capacitance value in each case?

9   Which is the larger capacitance: a 0 0001 mF or a 470 nF?

10   What is a 6800 pF capacitance in nF?

11   To increase a capacitance, would you add another capacitor in series or in parallel with the first?

12   Two inductors of 100 mH and 0.01 μH are connected first in series and then in parallel What is the total inductance value in each case?

13   To increase an inductance, would you add another inductor in series or in parallel with the first?

14   A current of 250 mA flows into a two-resistor parallel circuit. One resistor is 15 kΩ, the other is 25 kΩ Find the current flow in each.

15   In question 14, would each resistor have the same voltage measured across it?

16   A 50 V supply is connected across a series network of three resistors 12 kΩ, 8 kΩ and 5 kΩ Use the voltage divider equation to find the voltage across two of them and then deduce the voltage across the third.

17   The internal resistance of two particular voltage supplies are 10 Ω and 150 kΩ. Which is the better voltage source?

18   If the output of a voltage supply is measured at 30 V before a 25 kΩ resistance is connected across it, and drops to 10 V on being connected, what is the internal resistance?

19   If the maximum possible output of a current supply is 2 A but it delivers only 100 mA when a 1 kΩ resistance is connected across it, what is its internal resistance?

20   Two power supplies are available. One has R_{int} = 200 kΩ and one has R_{int} = 0.01 Ω. Which is the better voltage source and the better current source?

Further reading

Floyd T L (1993) *Principles of electric circuits*, 4th edn, pp 24–51, 102–8, 496–508, 552–67, 635–6 Merrill, Columbus, Ohio

Horowitz P & Hill W (1987). *The Art of Electronics*, sections 1 01–11 Cambridge University Press, Cambridge, England

Basic principles for electronics

Aims and objectives

The main aim of this chapter is to introduce you to some simple circuit representations in common use and the idea of impedance matching and circuit loading for different conditions. We will also look at equivalent circuits and their use in simplifying circuit calculations. The chapter also aims to illustrate and define the more common types of waveform seen in electronics.

After studying the topics in this chapter you will be able to:

- *determine the Thevenin and Norton equivalents of simple circuits*
- *understand why Thevenin and Norton circuits are used*
- *understand and explain what voltage, current and power transfer mean*
- *recognize the many different names used to describe internal and load resistances*
- *choose a load circuit to maximize for either voltage, current or power transfer*
- *recognize, draw and define the main types of voltage waveform in common use*
- *specify sine waves in terms of their r.m s., peak and d.c. offset values*

In the first chapter we looked at real and ideal voltage and current sources. In electronics we often have to deal with circuits which provide us with a voltage or a current but do not consist merely of a simple voltage source with two output terminals. For instance, the audio outputs from a TV or hi-fi must be sources of voltage and current, but certainly have more circuitry inside them than a single voltage and a resistor. The actual circuitry providing us with the output voltage or current is rather complex and it would be much easier if we could find a way of modelling it so that we could treat it as a simple real voltage or current source This is the perfect introduction for a look at the work of Thevenin and Norton.

2.1 Thevenin's theorem

The gist of Thevenin's theorem is that *any* circuit network of linear elements can be replaced by an equivalent circuit consisting merely of a single voltage source V_{th} and a single series resistance R_{th}. The concept is quite simple. It is saying that any circuit giving us a voltage output across two terminals can be simplified to look like a simple real voltage source When we make the appropriate calculations for this equivalent circuit, we call the equivalent voltage source 'the Thevenin voltage V_{th}' and the equivalent single resistance 'the Thevenin resistance R_{th}'. This circuit is shown in Figure 2.1

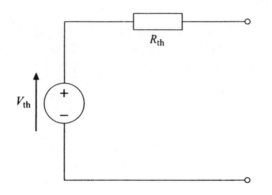

Figure 2.1

Obviously if we can simplify a circuit, we can simplify our analysis of it. This is the point: we are always looking for an easier way of doing things. While we are talking about making things easier, from now on we will use TEC to stand for Thevenin equivalent circuit. So how do we find the values for the TEC?

To calculate the values we follow three simple steps:

1. Find the open-circuit output voltage V_{oc} (across the output terminals) of the actual circuit. (Open-circuit output voltage means that a load resistance must not be added to the output while this is being found.)
2. In the actual circuit, replace any voltage sources with a short circuit and any current sources with an open circuit. Redraw the circuit and determine the total resistance R_{total} now appearing across the output terminals.
3. Then

$$V_{oc} = V_{th} \text{ and } R_{total} = R_{th}$$

We now know the steps, so let us try out an example. We assume that we have the circuit diagram available for any actual circuit in question, otherwise we have to take it to the laboratory and start connecting things to it.

EXAMPLE 2.1

Determine the TEC for the circuit in Figure 2 2

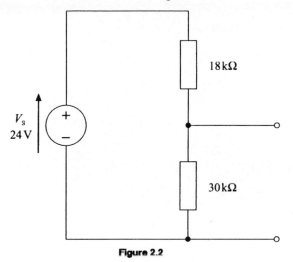

Figure 2.2

SOLUTION

This is a simple voltage divider and is often replaced by its TEC to simplify things in electronic analysis.

1   To find V_{oc} we simply use the voltage divider equation

$$V_{oc} = 24 \times 30 \text{ k}\Omega/(30 \text{ k}\Omega + 18 \text{ k}\Omega)$$
$$= 24 \times 30/48$$
$$= 15 \text{ V}$$

2.   Replace any voltage sources or current sources with short or open circuits respectively In this circuit we have one voltage source. We can therefore short it out and redraw the circuit as shown in Figure 2.3.

Notice that in Figure 2 3 we have also shown the effect of shorting the voltage source on the circuit layout. It has been redrawn to indicate that the two resistors now appear in parallel across the output terminals. Therefore

$$R_{th} = 18 \text{ k}\Omega \parallel 30 \text{ k}\Omega$$

$$= \frac{18 \text{ k}\Omega \times 30 \text{ k}\Omega}{18 \text{ k}\Omega + 30 \text{ k}\Omega}$$

$$= 11.25 \text{ k}\Omega$$

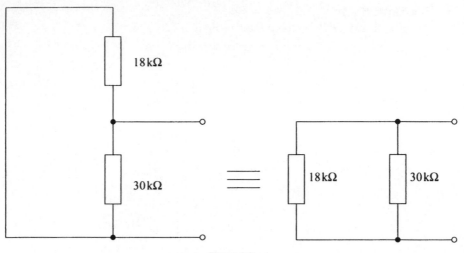

Figure 2.3

3   Now we can specify the TEC values

$$V_{th} = V_{oc}$$
$$R_{th} = R_{total} = 11.25 \text{ k}\Omega$$

The TEC is now as shown in Figure 2 4.

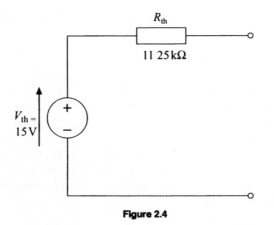

Figure 2.4

The TEC will tell us exactly how the actual circuit will perform when we add a load across the output terminals. If we add a specific R_L across the output of the real circuit, it will behave as if we had added the same R_L across the TEC. Figure 2.5 shows this. The important point in all this is that analysing the TEC is quicker than analysing the actual circuit. Whatever the open-circuit (terminal) voltage is, it

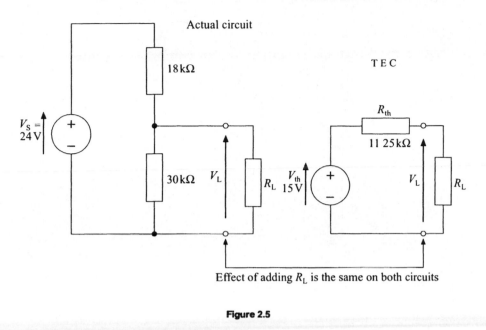

Effect of adding R_L is the same on both circuits

Figure 2.5

will drop when it is loaded by another circuit (Remember the case study in Chapter 1?) Finding out how much it drops is what makes the TEC so useful

Note that the TEC looks exactly like the representation of a real voltage source shown in Chapter 1. This is no coincidence: it is because of the usefulness of the TEC that real voltage sources are drawn in this way. In fact, the Thevenin resistance R_{th} and the internal resistance R_{int} are two (of many) different names for the same thing. In other words, the TEC tells us whether the output voltage comes from a good (low internal resistance) or a bad (high internal resistance) voltage source.

EXAMPLE 2.2
Find the TEC of the circuit in Figure 2.6

SOLUTION
1   We must first find the open-circuit output voltage V_{oc} (the output voltage when no current is drawn from the circuit, i.e no load is connected). To find the voltage at A (V_A) we use the voltage divider equation on V_s and the 12 kΩ and 25 kΩ resistances

$$V_A = \frac{24 \text{ V} \times 25 \text{ k}\Omega}{25 \text{ k}\Omega + 12 \text{ k}\Omega}$$

$$= 16.2 \text{ V}$$

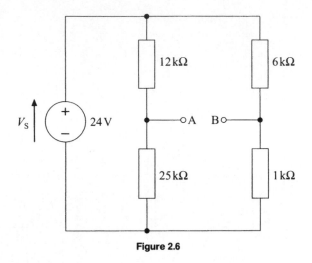

Figure 2.6

Similarly, the volts at B (V_B) are

$$V_B = \frac{24 \text{ V} \times 1 \text{ k}\Omega}{1 \text{ k}\Omega + 6 \text{ k}\Omega}$$

$$= 3 4 \text{ V}$$

Therefore

$$V_{oc} = (V_A - V_B)$$

$$= 12 8 \text{ V}$$

2   Replace any voltage sources by a short circuit and any current sources by an open circuit, and then redraw the circuit. (This is the hardest part for a circuit like this) Occasionally, more than one circuit redraw helps (see Figure 2 7)

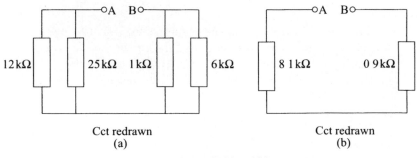

Cct redrawn
(a)

Cct redrawn
(b)

Figure 2.7 (a) and (b)

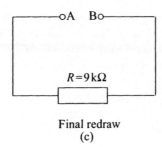

Final redraw
(c)

Figure 2.7(c)

We have shown that the total resistance across the output terminals has simplified down to 9 kΩ. Therefore

$$R_{total} = 9 \text{ k}\Omega$$

3   Now

$$V_{th} = V_{oc}$$

Therefore

$$V_{th} = 12.8 \text{ V}$$

$$R_{total} = R_{th}$$

$$= 9 \text{ k}\Omega$$

Finally the TEC in Figure 2 8 is drawn

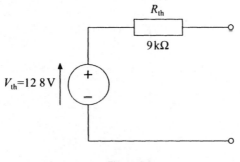

Figure 2.8

Study Example 2 2 to make sure you understand it If you have any problems, they are probably in understanding the redrawn circuits Many students find this difficult The following might help:

- The circuit in Figure 2.7(a) is the original circuit with a short circuit replacing the voltage source The tops of the 12 kΩ and 6 kΩ resistances are now connected directly to the bottoms of the 25 kΩ and 1 kΩ resistances. The circuit is drawn to show this more obviously. You can now see that the 12 kΩ and 25 kΩ resistances are connected in parallel (as are the 6 kΩ and 1 kΩ resistances).
- The circuit in Figure 2.7(b) just replaces the four values with the two new calculated parallel combinations of resistances

$$12 \text{ k}\Omega \parallel 25 \text{ k}\Omega = 8.1 \text{ k}\Omega$$
$$6 \text{ k}\Omega \parallel 1 \text{ k}\Omega = 0.9 \text{ k}\Omega$$

After this circuit has been drawn, it is easy to see that the two remaining values are series connected.
- The circuit in Figure 2.7(c) combines the series values to produce a single (8.1 kΩ + 0.9 kΩ) resistance of 9 kΩ.

Example 2.2. is now taken one step further to demonstrate its use in analysis.

EXAMPLE 2.3
Take the original circuit of Example 2 2 and imagine a load resistance of 5 kΩ being connected across the output terminals (see Figure 2.9). Now, what is the voltage across the 5 kΩ resistor?

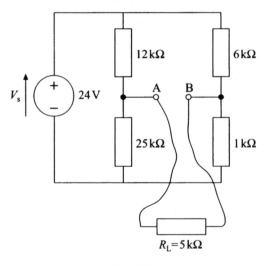

Figure 2.9

SOLUTION
This would normally mean going right back to the start and doing more circuit analysis. We could of course build it in the laboratory and measure

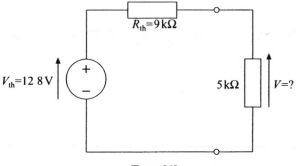

Figure 2.10

it or, and this is a good idea, we could simulate the circuit action using a computer software package. So how does a TEC help?

Simple We know what the TEC for this circuit looks like, so we draw it, attach the load resistance at the TEC output and apply the voltage divider equation. Figure 2.10 demonstrates this

$$V = (12.8 \times 5000)/14\,000$$
$$= 4.6 \text{ V}$$

This nicely demonstrates how much easier the TEC makes the calculation. In fact, you can now easily predict the output voltage of the circuit for any load resistance.

Self-assessment questions

SAQ2.1 When we are considering resistance values only, why do you think we replace a voltage source with a short and a current source with an open circuit? (Think ideally.)

SAQ2.2 If we removed the 5 kΩ load from the circuit in Figure 2 9 and replaced it with a 1 kΩ load, what would be the new load voltage?

SAQ2.3 Is the circuit shown in Figure 2.6 a good voltage source? Explain your answer

SAQ2.4 Using the circuit in Figure 2 2 and the associated TEC example, determine the new output voltage of the circuit if an 11.25 kΩ load were added

SAQ2.5 Repeat SAQ2 4 with a load of 50 kΩ.

SAQ2.6 Is the circuit in Figure 2.2 a good or a poor voltage source?

SAQ2.7 Draw a two-resistance voltage divider circuit with an input of 15 V and resistances of 22 kΩ and 48 kΩ. Show the output across the 48 kΩ resistance. Find the TEC and determine the output voltage when the following loads are connected across the output:
1.   3 kΩ
2   10 kΩ
3.   48 kΩ
4.   200 kΩ

2.2 Norton's theorem

We have looked at circuits which provide us with a voltage and the way they can be simplified by finding their TEC. Let us now consider circuits which provide us with a current output. This means that we need to use a simplified equivalent circuit with a current source representation. This is where the Norton equivalent circuit (NEC) comes in.

Norton's theorem tells us that any complex circuit of linear elements can be simply represented by a single current source and a single parallel resistance (Remember the real current source representation?) A typical NEC looks like the circuit in Figure 2.11. I_{Norton} is the Norton equivalent current source and R_{Norton} is the Norton equivalent parallel resistance.

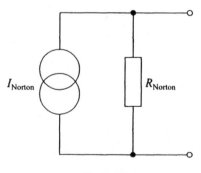

Figure 2.11

Our starting point is to assume that we have the circuit diagram of the original circuit with two output terminals and we want to find the NEC for it. We find the NEC by a method similar to that used to find the TEC. Look at the following steps:

1.   Determine the current that would flow through a short circuit connected across the output terminals This is the short-circuit current flow I_{sc}.
2.   In the actual circuit, replace any voltage sources with a short circuit and any

current sources with an open circuit Redraw the circuit and determine the total resistance R_{total} now appearing across the output terminals (just as in the TEC method)

3   Now we say

$$I_{\text{sc}} = I_{\text{Norton}}$$

$$R_{\text{total}} = R_{\text{Norton}}$$

That is all there is to it.

To highlight the links between an NEC and a TEC, we will look at the same circuit as the one we used in the TEC example

EXAMPLE 2.4
Determine the NEC for the circuit in Figure 2.2

SOLUTION
1   We draw the short circuit across the output and calculate the I_{sc} value This is shown in Figure 2.12 The 30 kΩ resistance is no longer part of the calculation. From Ohm's law

$$I_{\text{sc}} = V_{\text{s}}/18 \text{ k}\Omega$$

$$= 24 \text{ V}/18 \text{ k}\Omega$$

$$= 1.33 \text{ mA}$$

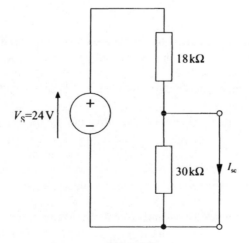

Figure 2.12

2.   Replace any voltage sources or current sources with short or open circuits respectively. This step is identical to step 2 in the TEC method We will therefore skip it and use the previous answer, which is

$$R_{total} = 11.25 \text{ k}\Omega$$

3   Now we say

$$I_{sc} = I_{Norton}$$

$$= 1.33 \text{ mA}$$

$$R_{total} = R_{Norton}$$

$$= 11.25 \text{ k}\Omega$$

The NEC is now as shown in Figure 2.13 Note that R_{Norton} is the same as R_{th} found earlier for the same circuit.

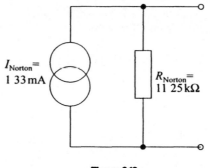

Figure 2.13

The NEC can be used to analyse the loading effects on the actual circuit just like the TEC examples earlier, except that we would probably be interested in the load current values rather than the load voltages. We would, of course, use the current divider equation, or just Ohm's law, to find the load currents resulting from different load resistances.

In respect of the relationship between the TEC and the NEC, we should note the following:

1.   A circuit can be simplified by finding either the Thevenin or the Norton equivalents. Which equivalent you choose is dependent only on whether you want the original circuit to act as a voltage source or a current source.

2   The Thevenin and Norton resistances of a circuit are equal.

3   The voltage developed across the R_{Norton} by the I_{Norton} is equal to the V_{th}, i e

$$R_{Norton} \times I_{Norton} = V_{th}$$

4.   Since $R_{Norton} = R_{th}$, we can also see that

$$I_{Norton} = V_{th}/R_{th}$$

This means that if one of the equivalent circuits has been determined, it is an easy job to find the other.

Self-assessment questions

SAQ2.8 Use the NEC in Figure 2.2 to determine the load current and voltage for load resistances connected across the output of:
1   $11\,25\,k\Omega$
2   $50\,k\Omega$

SAQ2.9
1   Find the NEC for the circuit in Figure 2.14.
2   Show that the NEC gives the same output voltage as the actual circuit when it is loaded by a $5\,k\Omega$ resistor.

SAQ2.10 Find the NEC for the circuit in Figure 2 15 Is it the same as the NEC in SAQ2 9?

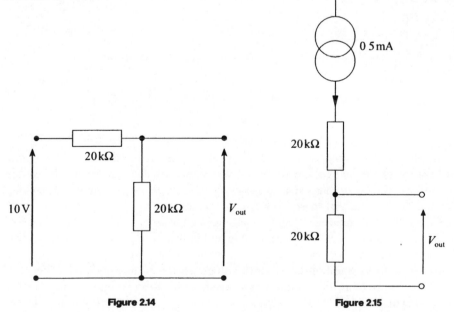

Figure 2.14   Figure 2.15

2.3 Impedance matching for voltage, current and power transfer

The major factor in understanding impedance (or resistance) matching stems from the realization that.

1. The output of almost any circuit can be represented as either a TEC or an NEC, where the associated internal resistance is more commonly called 'the output resistance' or 'source resistance'. Therefore, from now on the terms Thevenin resistance (R_{th}), internal resistance (R_{int}), output resistance (R_{out}) and source resistance (R_s) all mean the same thing.
2. The load circuit connected to the output of almost any circuit can be represented by a single resistance and is usually called either 'the load resistance' or 'the input resistance'. Therefore, from now on the terms load resistance (R_L) and input resistance (R_{in}) mean the same thing. Figure 2 16 shows the general case and all the names used to describe the resistances involved.

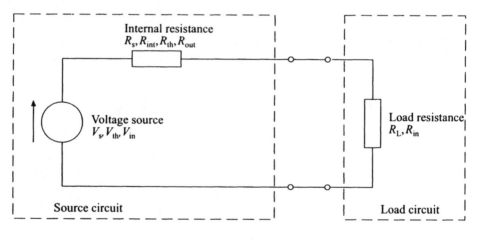

Figure 2.16

Impedance matching simply means finding the best values for the output impedance of the source and the input impedance of the load to satisfy certain conditions. The conditions we need to satisfy are voltage, current or power transfer. We will look at each of these by way of explanation.

Generally, the outputs and inputs of circuits are not purely resistive. They are very often combinations of resistive, capacitive and inductive elements (refer to Section 1.6 on impedances). For this reason, it is often useful to refer to their impedance values rather than simply the resistance values. However, to enable the concept of 'matching' to be more easily understood, we are going to stick with treating them as pure resistance values. For our purposes, therefore, whenever the word 'impedance' appears, we can replace it with 'resistance' (and vice versa). You can then start to consider the impedance values once you have a grasp of the ideas involved.

2.3.1 Impedance matching for voltage transfer

Voltage transfer means transferring a voltage from a source circuit to a load circuit. We usually want as much voltage as possible to be transferred and, when this has been achieved, we say that we have 'good voltage transfer'. For us to investigate the condition for good voltage transfer, we must look at the circuit representing our task Refer, therefore, to Figure 2.17 to see how voltage transfer is achieved.

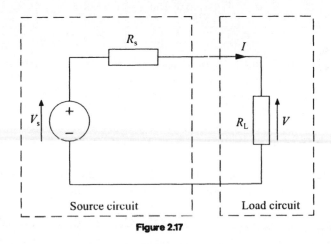

Figure 2.17

Using the voltage divider equation, we can say

$$V = V_s \times R_L/(R_L + R_s)$$

From simple mathematics we can see that if $R_s \ll R_L$, then R_s becomes negligibly small, giving

$$V \approx V_s \times R_L/R_L$$

$$\approx V_s$$

This tells us that if $R_s \ll R_L$, we get almost all of the source voltage V_s transferred to the load R_L. Therefore, we say for the general case that

the condition for good voltage transfer is met if $R_s \ll R_L$

A general rule is often used here. We have acceptable voltage transfer if R_L is at least ten times bigger than R_s. Let us try an example to convince ourselves and to get the feel of it

EXAMPLE 2.5

With reference to Figure 2.17, assume that we have a 6 V source with $R_s =$ 500 Ω, which is to drive a load circuit of $R_L = 100$ Ω:
1.   Do we have good voltage transfer?
2   What is the voltage across the load?
3.   How could we improve the voltage transfer?

SOLUTION
Using the voltage divider equation

$$V = 6 \times 100/(500 + 100)$$

$$= 1 \text{ V}$$

1.   Most of the voltage is dropped across the internal resistance of the source. We therefore have poor voltage transfer.
2.   The voltage across the load is only 1 V.
3.   We could improve the transfer either by increasing R_L (e g make $R_L = 10 \times 100 = 1$ kΩ) or by decreasing R_s (e.g. make $R_s = 500/10 =$ 50 Ω). This comes from our general rule for voltage transfer!
    To complete this example, let us take one of these improving suggestions and see what difference there is in the voltage transfer Try $R_L = 1$ kΩ. Then, again using the voltage divider equation

$$V = 6 \times 1000/(500 + 1000)$$

$$= 6000/1500$$

$$= 4 \text{ V}$$

This is a considerable improvement!

Working through Example 2.5, you can see that we get maximum voltage transfer when the input resistance of a load circuit is much larger than the output resistance of the source circuit. This can be achieved either by reducing the output resistance value or by increasing the load resistance value In real circuits we can usually arrange for one of these in order to achieve good voltage transfer

Self-assessment questions

SAQ2.11 The output impedance of a standard crystal microphone is 50 MΩ If it were connected to an amplifier of input impedance 20 kΩ, would we have good voltage transfer?

SAQ2.12 If the microphone in SAQ2.11 had a 500 mV$_{rms}$ output, how much of the signal would reach the amplifier unit?

SAQ2.13 Explain why a good voltage source will give better voltage transfer than a bad one. Give an example using some values of your own.

2.3.2 Impedance matching for current transfer

There are times when we are interested in transferring current rather than voltage These are when we have a current source (or at least a circuit that can be represented as a current source) which has connected to it a load circuit requiring as much current as possible from the source.

Current transfer means transferring a current from a source circuit to a load circuit. We usually want as much current as possible to be transferred. When this has been achieved we say that we have 'good current transfer'. To investigate the condition for good current transfer, we must look at the circuit representing our task, as we did earlier for the voltage transfer. Figure 2.18 shows how current transfer is achieved

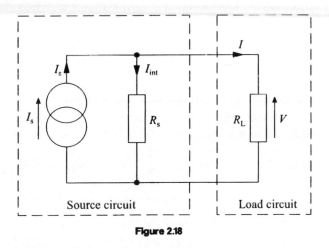

Figure 2.18

Using the current divider equation, we can say

$$I = I_s \times R_s/(R_s + R_L)$$

From simple mathematics we can see that if $R_L \ll R_s$, then R_L becomes negligibly small, giving

$$I \approx I_s \times R_s/R_s$$

$$\approx I_s$$

This tells us that if $R_L \ll R_s$, we get almost all of the source current I_s transferred to the load R_L. Therefore, we say for the general case that

the condition for good current transfer is met if $R_L \ll R_s$

A general rule is often used here We have acceptable current transfer if R_L is at least ten times smaller than R_s

Instead of working through an example, it might be more instructive to leave you to work through some questions on this yourself.

To summarize, you can see that we get maximum current transfer when the input resistance of a load circuit is much smaller than the output resistance of the source circuit. This can be achieved either by increasing the output resistance value or by reducing the load resistance value. In real circuits we can usually arrange for one of these in order to achieve good current transfer

Self-assessment questions

SAQ2.14 A good current source has a .. . . internal resistance value

SAQ2.15 Work through this example for yourself to help you remember current transfer. Draw a 0.1 mA current source with a 100 kΩ output (internal) resistance. Pretend for a moment you are experimenting to find the best load to add to it, to ensure lots of load current flow:
1.   Connect to it an 80 kΩ load. How much of the source current reaches the load?
2.   Then try a 400 kΩ load What is the load current?
3.   Finally, fit a 25 Ω load Is this any good? What is the load current this time?

Hopefully you have convinced yourself of the condition for good current transfer by working out the answers to these questions.

2.3.3 Impedance matching for power transfer

Impedance matching for power transfer may be the most important condition for applications involving output to the real world! It tackles the problem of transferring power from one circuit to another. Recall what was said earlier. that the output of a circuit can be represented as a voltage source or a current source A voltage source is fine if the load wants voltage. A current source is fine if the load wants current, but what if the load needs the maximum power available from the source? How do we represent the source to enable us to analyse the power transfer? The answer is that it does not matter which source we use: the outcome is the same.

If we use a voltage source circuit as shown in Figure 2.19, what size must R_L be in order to achieve maximum power transfer? The answer is that for maximum power transfer we need

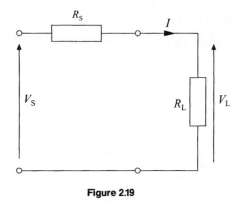

Figure 2.19

$$R_s = R_L$$

(Note the difference between this and the voltage and current transfer conditions.) This outcome is not obvious, but can be deduced intuitively by considering that for maximum V_L we want $R_L \gg R_s$ and for maximum I we want $R_L \ll R_s$, but because

$$\text{power } P = \text{voltage} \times \text{current}$$

we require a condition which satisfies both inequalities. Therefore, not surprisingly, for maximum power transfer

$$R_L = R_s$$

The proof for this condition has not been shown, although if your knowledge of calculus is reasonable, you could try the next exercise to prove that it is true To start with, in Figure 2.20 an experimental curve for the relationship between output power (P_{out}) at different values of R_L shows a distinct maximum $P_{out(max)}$ at a specific value of R_L. From the circuit diagram in Figure 2.19, we can say

$$P_{out} = I^2 \times R_L$$

where $I = V_s/(R_s + R_L)$ Therefore

$$P_{out} = [V_s/(R_s + R_L)]^2 \times R_L$$

Now, you can see from Figure 2 20 that the rate of change of output power with R_L has a specific value when the power reaches a maximum By differentiating you should be able to prove that, at maximum power

$$R_s = R_L$$

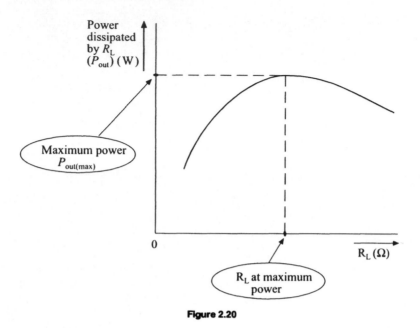

Figure 2.20

Self-assessment question

SAQ2.16 Try to prove this result using calculus

An obvious application of this condition occurs when you buy a new loudspeaker as a load for your audio amplifier. Look at the back of the hi-fi unit and you will see that a value is quoted for the R_s (or R_{out}) of the amplifier. (It is usually called output impedance rather than output resistance.) A typical value of this might be 8 Ω The transfer condition you are trying to optimize is for maximum power, and you will therefore be advised to buy a loudspeaker with a resistance of 8 Ω ($R_s = R_L$).

Self-assessment questions

SAQ2.17 A certain waveform generator has an output resistance of 600 Ω Choose a suitable load resistance to ensure:
1. good voltage transfer;
2. good current transfer;
3. good power transfer.
(Remember that zero and infinity are not sensible load resistance values for real circuits.)

SAQ2.18 Explain why the input resistance of a voltmeter needs to be very high

SAQ2.19 The output of a circuit is measured with a digital voltmeter (DVM) before and after adding a load resistance. (We can assume that the DVM input resistance is about $10^8\Omega$) The DVM reading drops by 50 per cent when a load of 1 kΩ is added What can we deduce about the circuit?

2.4 Voltage waveforms

It would be wrong to leave this chapter on basic principles without looking briefly at the sorts of waveform you are likely to come across on your electronics trail. Some of the waveforms shown are not explained in detail; they are included merely to inform you that they exist and that you may see more of them if you pursue your interest in electronics. However, we will spend a little time on the main ones, like the sine and square waves

2.4.1 The sine wave

The sine wave is the wave you will see and use until it is coming out of your ears, so let us describe it You may know most of this. skip it if you do. A voltage sine wave is usually described mathematically as

$$v(t) = V_p \sin(\omega t)$$

$v(t)$ is the instantaneous voltage value of the sine wave at angle ωt, V_p is the peak voltage or maximum value and $\omega = 2\pi f$, where f is frequency in hertz and t is time in seconds. (ωt) is in radians from 0 to 2π inclusive.

Since $\sin(\omega t)$ takes all values between -1 and $+1$ inclusive, multiplying this by V_p gives us the instantaneous voltage, which is anything from $-V_p$ to $+V_p$ inclusive.

To convert the radian axis to a time axis, we must remember that the period T is equal to $1/f$, so we can write

$$v(t) = V_p \sin[2\pi(t/T)]$$

This tells us that (t/T) is a fraction of the period at a time t. To find the radian value at this time, we multiply this fraction by 2π (since 2π is the number of radians in one full period) to give $2\pi(t/T)$ If the sine wave does not cross 0 radians at 0 V (which is normally assumed), the instantaneous value at time t is shifted by the amount the starting point shifts from 0 radians (see Figure 2.21)

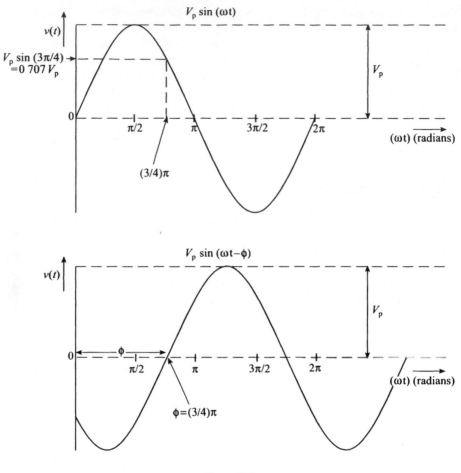

Figure 2.21

The difference in V_{rms}, V_p and V_{p-p}

It is not the intention to give a detailed account here· just the things you need to know for waveform sketching and description. First, the names.

- V_{rms} or root mean square voltage is the equivalent d.c. voltage value that would cause the same heating effect as the sine wave voltage
- V_p or peak voltage is the maximum voltage (or amplitude) in either the positive or negative direction. We have already looked at this term.
- V_{p-p} or peak to peak voltage is the maximum voltage measured between the positive and negative peaks

With reference to Figure 2.22

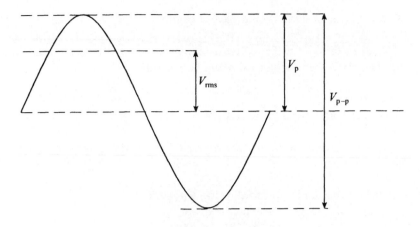

Figure 2.22

$$V_{rms} = V_p/2^{\frac{1}{2}}$$

$$= V_p \times 0.707$$

$$V_{p-p} = 2 \times V_p$$

$$V_p = V_{rms} \times 2^{\frac{1}{2}}$$

$$= V_{rms} \times 1.414$$

Note that V_p is sometimes written as V_{pk} or V_{peak} and V_{p-p} is sometimes written as V_{pk-pk} or $V_{peak-peak}$ Because of this, the following all describe the same sine wave

$$V_{pk} = 10 \text{ V}$$

$$V_{rms} = 7.1 \text{ V}$$

$$V_{p-p} = 20 \text{ V}$$

A word of advice· in case you ever find yourself forgetting whether to multiply or divide by $2^{\frac{1}{2}}$, all you have to remember is that the r.m s voltage is always less than the peak voltage. Since $2^{\frac{1}{2}} = 1.414$, you should be able to work out which operation is needed.

D.c. offsets
The usual centre voltage (or average voltage) of a sine wave is zero volts. If, however, the sine wave has a centre voltage different from zero, this new centre voltage is called a 'd.c offset' A d c. offset can be positive or negative A sine wave might therefore be described as a '100 Hz, $10\,V_{pk}$ sine wave with +6 V d.c.

offset'. This sine wave would have a centre voltage of +6 V and would oscillate from −4 V to +16 V.

You can think of a d.c offset as being the amount by which the sine wave is lifted up above zero or pushed down below zero.

Self-assessment questions

SAQ2.20 Draw the wave described as a 100 Hz, 10 V_{pk} sine wave with +6 V d.c. offset.

SAQ2.21 Draw and fully label a sine wave of:
1   V_{rms} = 240 V at 50 Hz with no d.c. offset;
2. V_{rms} = 12 V at 15 kHz and −15 V d.c. offset.

2.4.2 The square wave

There is little about the square wave to describe, except to refer you to Figure 2 23 The wave is rectangular, periodic (repetitive) and is defined by its frequency, d.c offset and amplitude In digital electronics we often see square waves with positive d.c. offsets lifting them up to sit on the 0 V line. It is traditional to describe a square wave in terms of its V_{p-p} value, e.g. a 5 V square wave usually means $V_{p-p} = 5$ V. If there is any doubt about the voltage value, a waveform sketch is the best (and quickest) way to describe it.

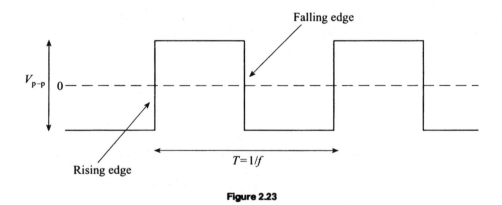

Figure 2.23

Note the rising and falling edges of the wave. The ideal square wave takes zero time to rise or fall. Therefore, an oscilloscope display of a good square wave will look like Figure 2.24(a), where the time taken for the electron beam to change voltages is too short for it to activate the phosphor screen A poor square wave looks like Figure 2.24(b), where the rise and fall times can clearly be seen

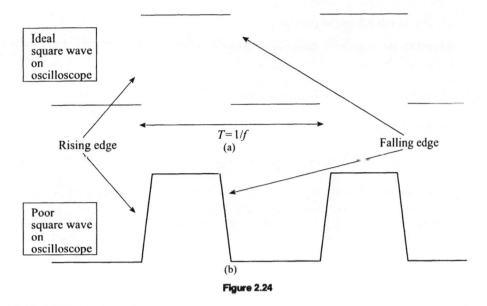

Figure 2.24

Because of the rapid changes in voltage levels, square waves are often used to test the frequency response of electronic equipment. Equipment not responding quickly enough to the wave tends to distort and smooth off the corners of the wave

2.4.3 The triangle wave

Again, the triangle waveform is better illustrated than described (see Figure 2 25) This waveform is commonly available on most waveform generators used in laboratories and workshops It is triangular in shape, periodic and is defined by its frequency, d.c. offset and amplitude. It is usually found with no d.c. offset and, because of its constant positive and negative gradients, it is used to test circuits for linearity of response

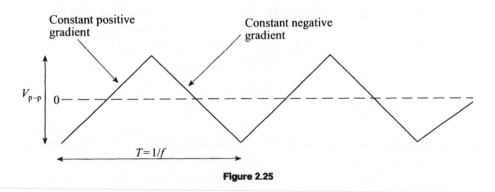

Figure 2.25

2.4.4 The sawtooth wave (ramp)

Looking like the edge of a saw, the sawtooth wave is not symmetrical about the voltage axis like the previous three waves are. It is more commonly called the 'ramp' wave (for obvious reasons). Often defined by its frequency and amplitude, it can, however, be non-periodic (non-repetitive), a single ramp being triggered by some action or other It is traditional to describe a sawtooth wave in terms of its V_{p-p} value, e g.

$$\text{a 5 V ramp wave usually means } V_{p-p} = 5\text{ V}$$

It is used to sweep a set of measurements in one direction with a rapid turn-off time before restarting and for timing/delay circuitry. It is used in the timebase circuit of an oscilloscope to sweep the electron beam steadily across the screen, it can be found in device testers and can form part of a delay timer. Figure 2 26(a) and 2.26(b) show the periodic and non-periodic versions respectively.

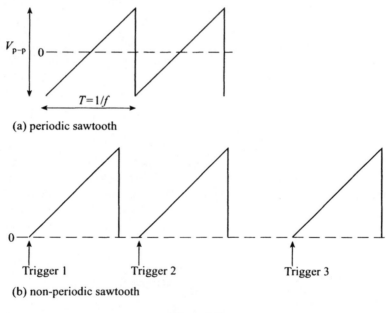

(a) periodic sawtooth

(b) non-periodic sawtooth

Figure 2.26

2.4.5 The pulse wave

Look at Figure 2 27. Imagine a square wave with different times of high and low voltage, and we have a pulse wave Like the sawtooth wave, it is non-symmetrical about the x axis and, although often defined by its frequency and amplitude, it can exist as a non-periodic wave, e g a single pulse being triggered by some action or

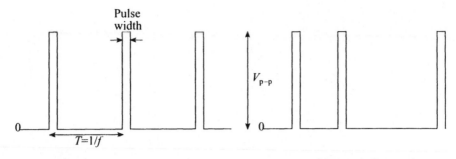

(a) Periodic duty cycle ≈ 15%   (b) Non-periodic one-shot

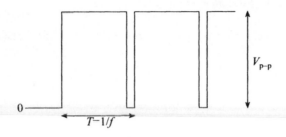

(c) Periodic duty cycle ~ 85%

Figure 2.27

other would produce a non-periodic wave In its non-periodic mode it is called a 'one-shot' or 'monostable' wave We see pulse waves and one-shot waves with d c. offsets (low voltage at zero) quite frequently

Other ways to describe the pulse wave are by pulse width, measured in seconds, or duty cycle, measured in percentage. Duty cycle is a measure of how much time, during one period, the voltage is at the high level Figure 2.27(c) shows a high-duty cycle

Self-assessment questions

SAQ2.22 Draw a 1 kHz square wave with a 10 V amplitude and +10 V d c offset

SAQ2.23 Describe the square wave in Figure 2 28

SAQ2.24 Draw a 10 kHz square wave with V_{p-p} = 10 V and +5 V d.c. offset.

SAQ2.25 A 5 kHz periodic pulse wave has a pulse width of 50 μs What is its duty cycle?

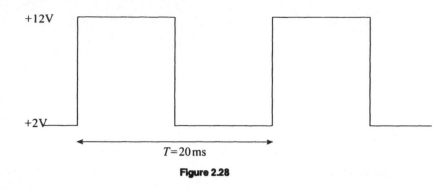

Figure 2.28

SAQ2.26 What is the duty cycle of a square wave?

SAQ2.27 Calculate the positive and negative gradient values of a 1 kHz triangle wave with a 5 V amplitude and zero d.c. offset. (Gradient is in volts per second Vs^{-1}.)

Checklist

Do you know:
- what TEC and NEC stand for?
- what a TEC and an NEC look like?
- the differences between source, output and Thevenin resistance?
- why TECs are useful in analysis?
- the difference between impedance and resistance?
- the meanings of voltage, current and power transfer?
- what load resistance to use for acceptable voltage transfer?
- the general rule for reasonable current transfer?
- the relationship between output and input resistances for maximum power transfer?
- how to determine a sine wave r.m s. voltage from a peak voltage?
- what d.c. offset means?
- what a sawtooth wave looks like?
- what duty cycle means?
- whether or not a pulse wave can be non-periodic?

Tutorial questions

1   Draw a two-resistance voltage divider circuit driven by a 35 V source. If the output is taken across a 3 9 kΩ resistor and the other resistor has 22 V dropped across it, find the TEC

2   Find the NEC for question 1.
3   If a load resistance of 1 5 kΩ were added across the output of the circuit in question 1, use the TEC to find the voltage across the load resistance
4   Draw a two-resistance voltage divider circuit with an input of 2 V and resistances of 20 5 Ω and 4.7 Ω Show the output across the 4 7 Ω resistance. Find the TEC and determine the output voltage when loads of 10 Ω, 1 kΩ, 15 Ω and 20 kΩ are connected across the output.
5   Draw a 0–10 V, 10 kHz square wave with duty cycles of 10% and 80%
6   Draw a 150 kHz square wave with a 5 V amplitude and +1 V d c offset
7   Draw a 500 Hz triangle wave with V_{p-p} = 1 V and +2 V d.c. offset.
8   A 100 kHz periodic pulse wave has a pulse width of 2 μs What is its duty cycle?
9   What is the pulse width of a 20 kHz square wave with a 20% duty cycle?
10 Calculate the positive and negative gradient values of a 15 kHz triangle wave with a 10 V d.c. offset.
11 Draw and fully label a sine wave of V_{rms} =12 V at 1.5 kHz and +6 V d c offset
12 Draw and fully label a sine wave of V_{rms} – 50 V at 10 Hz and   30 V d.c offset.
13 The output resistance of a certain power amplifier is 15 Ω Why would we choose to connect a 15 Ω loudspeaker to it?
14 Why does the input resistance of an ammeter need to be very low?
15 The output of a circuit is measured with a DVM before and after adding a load resistance. We can assume the DVM input resistance is about $10^8 \Omega$. The DVM reading drops to 10% of its original value when a load of 5 kΩ is added What can we deduce about the circuit?
16 A circuit has an internal resistance of 10 kΩ What would be a suitable load resistance to ensure good voltage transfer?
17 Do the Norton and Thevenin resistances of the same circuit always have the same value?
18 If a source with output resistance 2 kΩ is connected to a circuit of input resistance 20 kΩ, which would be the best transfer. voltage, current or power?
19 Describe a simple experiment using a variable resistance and a good voltmeter to find the internal resistance of a circuit.
20 Describe, as a mathematical expression, a 50 Hz, 20 V_{rms}, zero d c. offset sine wave.

Further reading

Floyd T L (1993) *Principles of electric circuits*, 4th edn, pp 286–317, Merrill, Columbus, Ohio
Horowitz P & Hill W (1987) *The art of electronics*, sections 1 05–11 Cambridge University Press, Cambridge, England

CHAPTER 3

Introduction to operational amplifiers

Aims and objectives

This chapter will help you to understand what an operational amplifier or 'op-amp' is and how it can be used Hopefully, it might convince you that designing circuits with operational amplifiers is easy! We will look at three of the basic amplifier circuits using op-amps, which will give you more insight into their usefulness (and their quirks).

You will see how these circuits form the backbone of many analog circuit designs. An understanding of their operation will provide you with a foundation for the analysis and application of many complex circuit designs, even at this early stage in your studies!

When you have studied this section you will have a better understanding of·

- *the general characteristics of an op-amp*
- *the basic differences between real and ideal op-amps*
- *what feedback is, and how it can be applied to an op-amp*
- *how to analyse an op-amp circuit*
- *how to determine circuit voltage gain and circuit input/output resistances*
- *the basic op-amp circuits, their names and their operation*
- *the design of op-amp inverting and non-inverting voltage amplifiers*
- *using the voltage follower as a buffer circuit*
- *how to select a circuit for a particular function*

3.1 The operational amplifier

Operational amplifiers or 'op-amps' are so called because they can be used to perform mathematical operations, such as addition, subtraction, multiplication, division, differentiation and integration, on input signals. Op-amps are being introduced at this early stage for three reasons:

1. They are easy to design with and to use.

2   They appear everywhere in electronics, taking the place of the more traditional transistor circuits

3.   To demonstrate the idea of treating a complex circuit as a 'black box'.

The first two reasons will become apparent to you as you learn more electronics The concept of treating something as a black box may not be familiar to you, so we will discuss this further.

The black box

We treat things as black boxes very often in our everyday life. Look at some of the electronic gadgets you have (or wish you could afford): a TV, hi-fi system or video recorder Do you know exactly how all the electronic circuitry in your TV works? Probably not But do you know how to use your TV? Do you know what it can do? Yes. The answer is yes because you treat the TV as a black box, i.e. you know little about its internal detail but you know how to use it and control it externally The name for the way a circuit connects with its surroundings is called its 'interface' A circuit interfaces with (connects with) its environment via a few input and output connections.

Self-assessment question

SAQ3.1 List the input and output connections to a standard portable stereo radio-cassette Label them as input or output

The point is that the *operation* of the gadget or circuit need not be known before you can *use* it to do something, and to make it do something you need to know how to connect it to the outside world. The term 'black box' is purely a theoretical concept and does not mean in any way that the object in question resembles a box, or is black! Strangely enough, in this world of TVs and integrated circuits, lots of things do *look* like black boxes Take the operational amplifier, for instance When you do your practical work, you will see that the op-amp is nothing more than a small black box with eight connectors We can use the op-amp by interfacing it to other circuitry via its eight connections (we usually use only five of the eight available) When we interface an op-amp to other circuitry, we need information about the connections to be made. The different pieces of interfacing information we need are called 'parameters' and they have names like input resistance (R_{in}), output resistance (R_{out}) and open loop voltage gain (A_{vol}).

The op-amp, when shown in a circuit diagram, has the circuit symbol with inputs and outputs as shown in Figure 3 1

Now we have mentioned the black box idea, we know that any circuit has inputs and outputs.

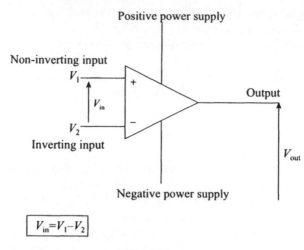

$$V_{in} = V_1 - V_2$$

Figure 3.1

Self-assessment question

SAQ3.2 Look at Figure 3 1. How many inputs and outputs does the op-amp have?

3.2 Describing an op-amp

An official description of an op-amp would read something like: 'it is a d.c.-coupled differential voltage amplifier, with a single-ended low resistance output, a very high input resistance and a very high voltage gain'. Let us break this down a bit.

D.c.-coupled means that the op-amp can be used with a.c. and d.c. input voltages. Differential amplifier means that it has two inputs. Single-ended output means that it has one output whose voltage is measured with respect to earth. High voltage gain basically means that the op-amp is a good voltage amplifier. Apply an input voltage (V_{in}) to it between its two input connections, and you will get an output voltage (V_{out}) which is much bigger. The number of times that its output is bigger than its input is called 'the open loop voltage gain' (A_{vol}) (v = voltage, ol = open loop) of the op-amp.

The open loop voltage gain of the op-amp is equal to the output voltage of the op-amp divided by the input voltage of the op-amp, i.e.

$$A_{vol} = V_{out}/V_{in}$$

A typical value of A_{vol} for an op-amp is in the region of 10^5–10^6. Rather large! This gain figure is dimensionless

Self-assessment questions

SAQ3.3 Look at the op-amp in Figure 3 1 If $V_{in} = 10\ \mu V$ and $V_{out} = 2\ V$, what is the value of A_{vol}?

SAQ3.4 What is the expression for V_{out} in terms of A_{vol} and V_{in}?

3.2.1 The inputs and outputs of the op-amp

We can represent any amplifier by the triangle symbol (shown in Figure 3.1), but for our black box representation we need to know a little more. We need to have a very simple way of showing how our input and output will respond to external connections. In simple terms, what does the op-amp input look like to an external circuit? The same thing can be asked of the op-amp output. You can already answer these questions if you think back to Section 2.3 on impedance matching.

You need to realize, of course, that the op-amp input looks like a load circuit to any circuit connected to its input.

Self-assessment question

SAQ3.5 If the op-amp input looks like a simple load circuit, how can it be represented? Draw the op-amp triangle symbol and have a go at adding the input load circuit representation.

You *should* have drawn a resistance in the answer to this question. This resistance is a load for the circuit connected to the op-amp input Because we are looking into the op-amp input, the name for this resistance is 'input resistance' (R_{in}).

Self-assessment question

SAQ3.6 If R_{in} is the load seen by any driving circuit connected to the op-amp circuit, what would be a good value for R_{in} to ensure good voltage transfer from the drive circuit to the op-amp input?

What about the op-amp output representation? If the op-amp is a voltage amplifier it can provide a voltage at its output (V_{out}). We can say that the op-amp is a source of voltage. Let us call it a voltage source. Even better, let us say that the output stage of the op-amp can be represented by its Thevenin equivalent circuit! You will know all about voltage sources and their associated internal resistances if you have read this book from Chapter 1. If you have not, then look back at Section

1.10 on voltage and current sources. The output of an ideal op-amp looks like an ideal voltage source. A real op-amp has a real voltage source as an output

Self-assessment question

SAQ3.7 What does this tell you about the size of the internal resistance value of the output of a real op-amp operating as a good voltage source?

One thing you should remember about this resistance value is that it can be called the internal resistance, the Thevenin resistance or the output resistance, each of which is correct, but it is usually called 'output resistance' (R_{out}).

If the value of the resistance (either input or output) is frequency dependent (changes with frequency), then it is not simply resistive: it must also have capacitive and inductive properties For this reason, output and input resistances are often combinations of resistive and reactive components, and are therefore called 'impedances' (Z) (Section 1.6). For example

$$\text{input impedance} = Z_{in}$$

This should be enough to enable you to go back to the op-amp symbol you drew and to answer the next question.

Self-assessment question

SAQ3.8 Try to add to the symbol a representation of the output circuit This is not quite so easy to get exactly right, but try before looking at the answer.

3.2.2 The symbol representing the op-amp

The way we represent the op-amp as a symbol depends on the context of the symbol. If we were trying to illustrate the R_{in} and R_{out} properties of an op-amp in a circuit, we would draw the symbol complete with its input and output representations, but normally a simple triangle will be sufficient to indicate the op-amp's presence in a circuit diagram.

Although the op-amp always requires a power supply connection to be made in the practical circuit, usually we do not show these connections on the symbol so that circuit diagrams are less cluttered.

Notice in Figure 3.1 that the inputs are labelled + (non-inverting) and − (inverting). The words clarify the meaning better than the + and − symbols. They refer simply to the phase difference between a voltage change at an input and its effect on the amplified voltage change on the output.

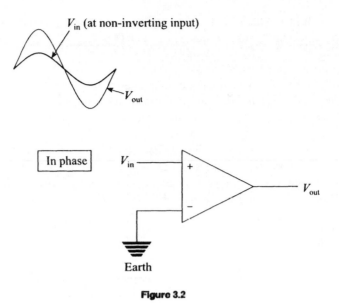

Figure 3.2

To illustrate this, consider adding a voltage signal at the non-inverting input as shown in Figure 3.2, while keeping the inverting input held at earth potential (0 V) Now, any voltage change on the non-inverting input will be amplified at the output, and the input and output signals will be in phase In other words, the output will be a non-inverted version of the input.

The input and output voltages are not drawn to scale in Figure 3.2 In reality, V_{in} would be tiny and V_{out} would be large because of the large A_{vol} of the op-amp

Given the information that the relationship between the inverting input and V_{out} is a 180° phase shift, imagine a case similar to that in Figure 3.2, but this time with the non-inverting input at earth and the inverting input having a sine wave

Self-assessment question

SAQ3.9 Draw what you think the input and output waveforms for the case above would look like (Remember to label the waveforms and show a small sketch of the circuit used, as is done in Figure 3 2)

3.2.3 Typical values

To complete this introduction to op-amps, it would be appropriate here to make a comment about real op-amp parameters. There are many parameters quoted when you delve into the specifications of a real device, but for starters we will take the R_{in}, R_{out} and A_{vol} values for a commonly used industry standard op-amp. the 741. Its full name includes a prefix which depends on the manufacturer, e.g. the uA741 made by Fairchild and the LM741 made by National. The 741 is a nice, easy, stable and, above all, cheap op-amp. It has $A_{vol} \simeq 200\,000$, $R_{in} \simeq 2\,\text{M}\Omega$ and $R_{out} \simeq 75\,\Omega$ These typical values are by no means exact (even in a batch of op-amps of the same type), e g for the standard 741 op-amp, we may often find that the A_{vol} may vary from 50 000 to 500 000. The input and output resistance values may also be a little unpredictable, although R_{in} is more likely to be known more accurately than A_{vol} We just hope that any op-amp performs as a high gain circuit with a large input resistance and a small output resistance.

There are many times when circuit characteristics, especially the gain value, must be known more accurately. The question is, how can we know the voltage gain of the circuit if the voltage gain of the particular op-amp is not accurately known? This is where feedback comes in.

3.3 Feedback

First of all, if there is a connection from an output to an input of a system, generally we have feedback The term comes from the idea that something is fed back to the input from the output which has an effect on the overall performance of the system. The operation is often called 'adding a feedback loop'. Look at Figure 3.3 to see a black box representation of using two types of feedback with an op-amp.

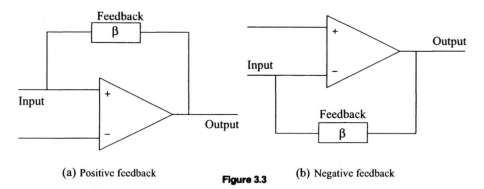

(a) Positive feedback   **Figure 3.3**   (b) Negative feedback

In electronics we can feed back currents or voltages, and we can choose the *type* and the *amount* of feedback to use.

3.3.1 The feedback fraction

We have a choice do we feed back *all* of the output voltage (100%) or only a little bit of it (say 1%)? The amount we use is called the 'feedback fraction' (β) Note that the value of β is always between 0 and 1, i.e.

$$0 < \beta \leq 1$$

3.3.2 Types of feedback

Feedback can be used to pick off some output voltage or current and reintroduce it at the input of the circuit. It is the *way* it is fed into the input that causes changes in circuit performance to occur, i.e it can reinforce (add to) the input signal or it can oppose (subtract from) the input signal. As the overall effects of the two types are very different, we will briefly look at each in turn.

If the feedback adds to the input signal it is called positive feedback and usually results in an output signal that is larger than it would have been without the feedback connection. Circuits using positive feedback generally have unstable states (or non-linearity), unpredictable parameters and are usually oscillators or electronic switches.

If the feedback subtracts from the input signal it is called negative feedback and results in an output signal that is smaller than it would have been without the feedback connection. Circuits using negative feedback are usually stable, predictable, linear circuits with well-defined parameters and characteristics

The use of negative feedback is extremely important and will be illustrated in many of the following op-amp circuits. In fact, using negative feedback, op-amps of widely varying A_{vol} values can be used in the same circuit to give *almost* the same result. Note that I avoid saying *exactly* the same result. You can see in Figure 3 3 the usual way of connecting up each type of feedback.

Read the whole of Section 3.3 on feedback again and then try to write in the missing words below

Self-assessment questions

SAQ3.10 If feedback is applied which .. .. from the input signal, it is called negative ...   .

SAQ3.11 The amount of feedback used is called the .... and has the symbol .... .

SAQ3.12 If the feedback to the input adds to the input signal, It is called .... feedback and causes the output to be .. . .

SAQ3.13 If the feedback signal subtracts from the input signal, it is called .. .. feedback and causes the output to be . ...

SAQ3.14 Oscillator circuits usually utilize . ..... feedback.

SAQ3.15 Stable, linear circuits usually utilize .. . .. feedback

3.4 Using op-amps in circuits

We are first going to consider a few basic op-amp circuits that you will see in regular practical use for many applications Generally, op-amp circuits fall into two categories

1. linear op-amp circuits,
2   non-linear op-amp circuits

In this chapter we will deal in detail with the three main linear amplifier circuits.

1   the inverting amplifier,
2   the non-inverting amplifier;
3. the voltage follower.

All these circuits use negative feedback
Simple op-amp circuit analysis on these circuits is so easy! We need to learn and apply two rules only. In our analysis, we consider that the op-amp has idealized parameters Think what this means. We know that a real op-amp has a very large A_{vol} value, a large R_{in} value and a small R_{out} value. Ideally, therefore, we would like both its A_{vol} and R_{in} to tend towards infinity. Although a real device never quite manages to attain perfection status, the amount it falls short of this usually makes no difference in the circuit action. So we can assume we are using an idealized op-amp Let us now look at how the two rules can be created.

Creating rule 1
Think about a real op-amp for a moment. In practical circuits, the op-amp output voltage (V_{out}) rarely exceeds about 15 V, which is an amplified version of the input voltage (V_{in}). If the A_{vol} of our op-amp is, say, 600 000, what size is our input voltage (V_{in})? (See Figure 3.4.)

$$V_{in} = V_{out}/A_{vol}$$
$$= 15 \text{ V}/600\,000$$
$$= 25\,\mu V$$

All we now say is that 25 μV is negligibly small and can be considered to be almost zero. Hence, the input voltage is almost equal to zero. Simple!

RULE 1: The voltage difference between the inputs is almost zero.

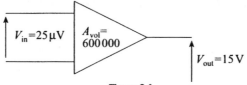

Figure 3.4

Creating rule 2

Think about real circuits again. Remember that a source circuit driving the op-amp will see the op-amp input as a load resistance of value R_{in}. How much current will flow into R_{in} if R_{in} is made very large? You will agree that as the value of R_{in} increases, the current flow through it must reduce (see Figure 3 5). We therefore reach a point where R_{in} is so large that the current flow is negligibly small. Let us call the current flow almost zero for simplicity. We now have rule 2 sorted out!

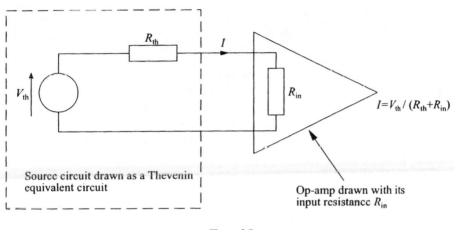

Figure 3.5

RULE 2· The op-amp inputs draw almost no current

Look at the two op-amp rules again, and remember them.

Before we move on, I would like to add one small word of caution. These rules apply to many op-amp circuits *except* when a particular condition occurs at the output, this is called 'saturation' If saturation occurs at the output, rule 1 may no longer be true. We will look at this condition later. For now let us see how we apply these rules

In all the following negative feedback circuits, the feedback connection is to the inverting input.

We have done enough theorizing. Let us now look at a real circuit.

3.4.1 The op-amp inverting amplifier

Look at the inverting amplifier in Figure 3.6 Notice the resistance R_2 connecting the output to the input. Its job is to feed back some current from the output terminal to the input terminal. But, you may ask, why is the current I_f shown flowing from input to output, rather than the other way round? The answer to this is twofold:

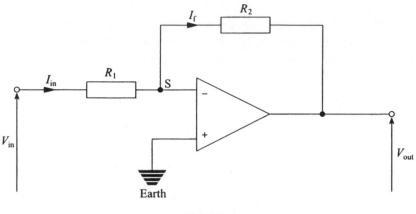

Figure 3.6

1. Applying feedback merely means making a connection from output to input to allow the feedback current to flow in either direction.
2. It does not really matter which way we show it drawn on the circuit diagram, provided that in the analysis we remember which end of the resistance is the more positive.

A little reminder here: the end into which the current flows is the positive end.

Analysis

So what are we going to analyse? For any circuit we are always interested in finding the parameters that describe the circuit performance. Remember the interface and black box idea from earlier on? Well, we now want to be able to take a complete circuit and treat it as a black box. We are therefore interested in the input and output resistances, and the voltage gain of the complete circuit. To enable us to indicate the difference in the circuit parameters and op-amp parameters, we will use the following terms. input resistance of the circuit (R_i), output resistance of the circuit (R_o) and voltage gain of the circuit ($A_v = V_{out}/V_{in}$) A_v is sometimes called the 'closed loop voltage gain' of the circuit because of the introduction of the feedback loop in the circuit.

Voltage gain

We merely apply our two rules to the circuit.

Applying rule 1, if the non-inverting input is held at earth potential, then the voltage at the inverting input must also be at earth potential. Therefore

$$\text{voltage at point S} = 0\,V$$

Applying rule 2, if the op-amp input draws almost no current, then the current I_{in} is equal to the current I_f

$$I_{in} = I_f$$

We now define the currents in terms of voltages and resistances. Remember point S is at 0 V Therefore

$$I_{in} = (V_{in} - 0)/R_1$$

$$= V_{in}/R_1$$

$$I_f = (0 - V_{out}/R_2$$

$$= -V_{out}/R_2$$

Equating these equations, we get

$$V_{in}/R_1 = -V_{out}/R_2$$

Rearranging this to find voltage gain, we get

$$V_{out}/V_{in} = -R_2/R_1$$

This means that

$$A_v = -R_2/R_1$$

It is important to note three things here:

1   The circuit voltage gain no longer involves any information concerning the op-amp, it is merely the ratio of the two external resistors R_2 and R_1.
2   The minus sign denotes an inverting action between the input and output.
3. The circuit will amplify d.c. and a.c voltages equally.

Let us look at an example to illustrate these points. Imagine you have built the circuit in Figure 3.7 The $A_v = -10$ kΩ/1 kΩ = 10; therefore, a 1 V_{pk} input sine wave will be amplified by 10 and inverted. (Note that inverted means that there will be a 180° phase shift in the output signal, as seen in Figure 3 7.)

What about a d.c. signal? A good question. At d.c. we obtain the same effect. An input of +1 V will emerge as a −10 V output. The inversion now merely means the output is of the opposite polarity.

Okay, that takes care of the voltage gain but, before moving on to find R_i, a word about the point S in the circuit of Figure 3.6, which is called a virtual earth point.

Virtual earth

You can see from the analysis that point S is at earth or 0 V potential, although it is not connected to the earth. (In fact, there is a considerable amount of circuitry between that point and the earth point) The point S is, in effect, an earth but not an

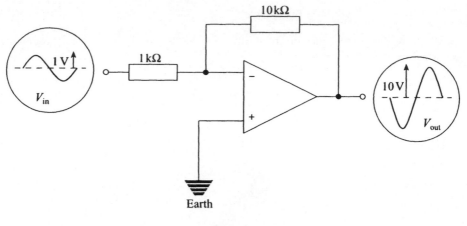

Figure 3.7

actual earth. Any point like this, in any circuit, is called 'a virtual earth'. If you are hankering for a definition, you could say that 'a virtual earth is a point in a circuit that is held at earth potential but is not connected to earth'

Input resistance of the inverting amplifier

What about the R_i value of the inverting amplifier in Figure 3 6? The input resistance of any circuit is determined from Ohm's law, i e

$$\text{input resistance} = \text{input voltage/input current}$$

For the inverter, the input voltage is V_{in} and the input current is I_{in}, but since point S is a virtual earth

$$I_{in} = (V_{in} - 0)/R_1$$
$$= V_{in}/R_1$$

Therefore

$$R_i = V_{in}/I_{in}$$

becomes

$$R_i = V_{in}/(V_{in}/R_1)$$

(V_{in} cancels.) Therefore

$$R_i = R_1$$

It is important to note two things here

1 The input resistance of the circuit no longer involves any information concerning the op-amp
2 The R_i value is simply the size of the external resistor R_1 in the input circuit

Self-assessment questions

SAQ3.16 What is the input resistance of the circuit in Figure 3.7?

SAQ3.17 What is the expression for the voltage gain of the inverting circuit?

SAQ3.18 There should be a minus sign in your answer to SAQ3 17. What do you think it means?

SAQ3.19 If $R_2 = R_1 = 1\,k\Omega$ and $V_{in} = 2\,V$, what is V_{out}?

SAQ3.20 If R_2 is increased to $4\,k\Omega$, what will V_{out} now be?

SAQ3.21 What is the input resistance (R_i) of the inverter circuit in the last two questions?

Output resistance

Finally, let us consider the output resistance R_o of the circuit The simple answer here is that R_o is very small. The reasoning is again twofold.

1 A real op-amp has a low output resistance anyway (look at the 741 typical value) as it is a voltage source.
2 If we apply feedback analysis (something to look at later), we find the *already* low R_o value is reduced considerably by the negative feedback action. We simply consider R_o to be negligibly small for most circuits with negative feedback.

Therefore, $R_o = 0\,\Omega$.

Self-assessment question

SAQ3.22 Design and draw the circuit of an op-amp inverter circuit with $R_i = 20\,k\Omega$ and $A_v = -20$.

3.4.2 The non-inverting amplifier

As its name implies, the non-inverting amplifier circuit will amplify an input signal, giving us an output which is a non-inverted version of the input. In other words, the output is in phase with the input. (This is obvious really if you remember that the input signal is fed directly to the non-inverting input.)

Analysis of the non-inverting amplifier

With reference to the circuit in Figure 3.8, we will apply our two rules to find A_v, R_i and R_o, just as we did for the inverting amplifier

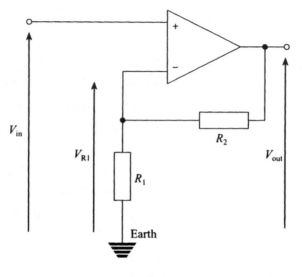

Figure 3.8

Voltage gain

As before, we will find that the gain A_v is determined purely from the external resistances. Before launching into the theory, note that this non-inverter is merely the inverter circuit with the earth and the V_{in} connections swapped round.

Applying rule 1, we get

$$V_{in} = V_{R1}$$

but by treating R_1 and R_2 as a potential divider between earth and V_{out}, we have

$$V_{R1} = V_{out} \times R_1/(R_1 + R_2)$$

Equating these equations, we get

$$V_{in} = V_{out} \times R_1/(R_1 + R_2)$$

Rearranging this gives

$$V_{out}/V_{in} = (R_1 + R_2)/R_1$$

Therefore

$$A_v = (R_1 + R_2)/R_1$$

We can remember this more easily if we write it as

$$A_v = (R_1/R_1) + (R_2/R_1)$$

which simplifies to

$$A_v = 1 + (R_2/R_1)$$

It is important to note five things here:

1. The circuit voltage gain no longer involves any information concerning the op-amp. It is merely the sum of 1 plus the ratio of the two external resistors R_2 and R_1.
2. There is no minus sign, and therefore no inverting action, between the input and output
3. The circuit will amplify d.c. and a.c. voltages equally
4. The expression is like that of the inverting amplifier with 1 added.
5 The circuit uses negative feedback, where R_2 is the feedback resistor

Self-assessment questions

SAQ3.23 Imagine we have a non-inverting circuit built with $R_2 = 10\,k\Omega$ and $R_1 = 2\,k\Omega$. What is the voltage gain of the circuit?

SAQ3.24 Another non-inverter circuit has $V_{in} = 500\,mV$, $R_2 = 5\,k\Omega$, $R_1 = 0.5\,k\Omega$ What is V_{out}?

Input resistance

What else is important about the non-inverter? One thing we can say is that it has a very predictable input resistance value. You will recall that to find the value of the input resistance of any circuit, we simply measure the input voltage and the input current, and then use Ohm's law

$$R_i = V_{in}/I_{in}$$

Let us do this for the non-inverting circuit. V_{in} is just whatever our input voltage is set at What about I_{in}? Remember rule 2? In this circuit, $I_{in} \approx 0$. This means, of course, that R_i for the ideal non-inverter is infinitely high! In fact, when we use a real op-amp (non-ideal), R_i is still found to be extremely large. Something we will look at later is that the negative feedback connection helps to raise the circuit input resistance to a value much higher than the actual input resistance of the op-amp. (It reaches anything from tens to hundreds of megohms!) The bit to remember is

$$R_i = \infty \ \Omega$$

Self-assessment question

SAQ3.25 Think about voltage transfer for a minute. If the input to a non-inverter were used as a load on a poor voltage source, what sort of voltage transfer would we have? Good, bad or indifferent? Justify or comment on your answer.

Output resistance

What size is R_o for the non-inverting circuit? The answer is the same as the answer for the inverting amplifier. We simply consider R_o to be negligibly small for most circuits with negative feedback. Therefore

$$R_o = 0 \ \Omega$$

As we will see later, the more detailed answer to this is that by applying negative feedback directly from the output to the input, we can show that the (already) low output resistance of the op-amp is reduced drastically to values of the order of small fractions of ohms (milliohms). Of course, this is good. It means that even non-ideal op-amp circuits (this is the real world we are in!) will give us output voltages that look very much as though they are coming from the ideal voltage sources. (Those of you having already studied feedback theory will realize that both the inverting and non-inverting circuits have low R_o values because they both utilize voltage-derived feedback loops.)

3.4.3 The op-amp voltage follower

It is a little unfair to treat the op-amp voltage follower as another new op-amp circuit, because it is really only a *special case* of the non-inverting circuit. Here is why. Look back at the non-inverting circuit (remember $A_v = 1 + (R_2/R_1)$) and imagine the case where we want to reduce its voltage gain by reducing R_2 and increasing R_1. If we continue in this fashion, we reach a theoretical limit where R_2 reaches zero and R_1 reaches infinity. This, of course, means that R_2 is a short circuit and R_1 is an open circuit. This situation is shown in the circuit diagram in Figure 3.9. It is called a 'voltage follower'

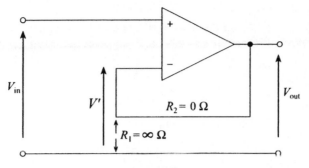

Figure 3.9

Voltage gain

Applying rule 1 again to Figure 3.9

$$V_{in} = V'$$

But

$$V' = V_{out}$$

Therefore

$$V_{in} = V_{out}$$

$$A_v = 1$$

There are four important points to note here:

1. The output is the same size as the input
2. The voltage gain is 1.
3. The output voltage will follow the input voltage exactly – hence its name.
4. As the circuit is a special case of the non-inverting circuit, the output is in phase with the input.

An alternative analysis for this circuit takes the non-inverting gain expression as its starting point. It gives

$$A_v = 1 + (R_2/R_1)$$

If R_2 is reduced to $0\ \Omega$ and R_1 is increased to infinity Ω, we have the new expression

$$A_v = 1 + (0/\infty)$$

which becomes

$$A_v = 1$$

Input resistance

The same argument holds for the input resistance of the voltage follower as for the non-inverter. We will therefore cut this analysis short and say directly

$$R_i = \infty \ \Omega$$

Output resistance

Again, nothing has changed from the R_o value of the previous circuits

$$R_o \approx 0 \ \Omega$$

Let us summarize the main points about the voltage follower:

1. V_{in} and V_{out} are the same size
2. V_{in} and V_{out} are in phase
3. R_i is very large (ideally, $R_i = \infty$).
4. R_o is very small (ideally, $R_o = 0$).

You would not be the first person to be mystified as to why the voltage follower is used at all. After all, it only seems to give out the same voltage as you put in, so what use is it?

Why the voltage follower is used

The usefulness of the voltage follower is in the vastly different input and output resistances it offers to source and load circuits, without changing the size of the voltage. In applications it often appears under different names, such as 'buffer circuit', 'voltage buffer' and 'impedance changer'. Look at the following example to see how the voltage follower is used to match a poor voltage source to a load.

Application of the voltage follower as a buffer circuit

Look at Figure 3.10, where we have a common electronics problem. A 1 V source circuit with $R_o = 1000 \ \Omega$ must be connected to a load circuit with $R_i = 10 \ \Omega$. Ideally, we want good voltage transfer. In other words, we want as much as possible of the 1 V to appear across the input of the load circuit. The reality of the situation is, unfortunately, rather different. Before closing the switch SW1, V equals 1 V (off-load, no current flow). When the source and load circuits are connected by closing SW1, potential divider theory tells us that the voltage reaching the load (V_L) is

$$V_L = 1 \times 10/(10 + 1000)$$

$$= 0.01 \ V$$

Virtually nothing reaches the load! As mentioned in Chapter 2, this is poor voltage transfer. How can we solve it? This is where the voltage follower comes into its own. We cannot change the resistance values, but we *can* change the loading effect.

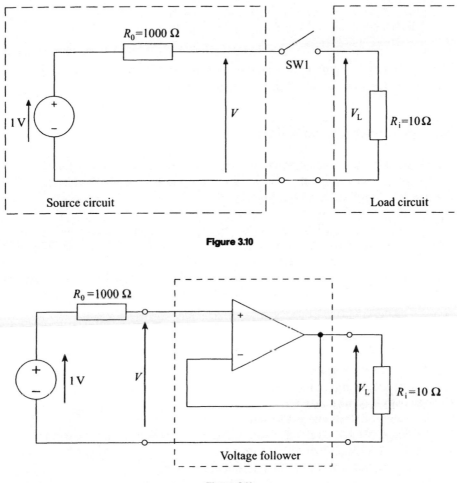

Figure 3.10

Figure 3.11

Now look at Figure 3 11. If we insert a voltage follower circuit between the two circuits, the follower's R_i equals infinity Therefore

$$V = 1 \times \infty/(1000 + \infty)$$
$$= 1 \text{ V}$$

A_v equals 1 and therefore

$$V_{out} = 1 \text{ V}$$

The voltage follower's R_o equals 0 Ω This gives the load voltage

$$V_L = 1 \times 10/(10 + 0)$$
$$= 1 \text{ V}$$

The circuit acts as a buffer between the two circuits and effectively transforms the poor voltage source into an ideal voltage source

Self-assessment question

SAQ3.26 Using the source and load resistance values given in Figure 3 10, work out the new voltage that would be developed across the load if the voltage follower had $R_i = 1$ MΩ and $R_o = 1$ Ω.

When we create circuits like the one in Figure 3.11 there are many resistance values to consider. To make matters worse, there are many ways of describing each one, e.g. source resistance, internal resistance, output resistance, load resistance and Thevenin equivalent resistance It is therefore probably instructive to try to answer the following questions where different terms are used. See how you get on.

Self-assessment question

SAQ3.27 Would the voltage transfer in SAQ3.26 be better or worse by making the following alterations.
1.   reducing R_i of the load circuit,
2.   reducing R_i of the voltage follower;
3.   making R_i of the load circuit 20 times bigger than R_o of the source circuit,
4.   making R_o of the follower circuit 1 kΩ;
5.   increasing the size of the voltage from the source;
6   increasing the source resistance of the source circuit;
7.   increasing the load resistance of the load circuit;
8.   reducing the Thevenin resistance of the 1 V source;
9   increasing the internal resistance of the voltage follower output.

These types of buffer circuit are often used between transducers (or sensors) and signal processing circuits (e.g. between microphones or tape heads and amplifiers) or any time when signal voltages come from poor (or, more likely, unpredictable or unknown) sources of voltage.

This completes the introduction to linear op-amp circuits but, before moving on to the next chapter, have a quick look at the final checklist

Checklist

Do you know:
- why the op-amp is so called?
- what real and ideal op-amp values of A_{vol}, R_{in} and R_{out} are?
- what the feedback fraction is, what its symbol is and what its range is?
- what type of feedback is applied in the inverter circuit?
- what the two rules for op-amp analysis are?
- how to draw the circuit diagram of an op-amp inverter?
- how to derive the A_v of the op-amp inverter?
- what a virtual earth is?
- what the approximate R_o of the op-amp inverter is?
- how to draw the non-inverting amplifier circuit diagram?
- how to derive the A_v expression for the non-inverting amplifier?
- how to design a non-inverting amplifier with a gain of +5?
- the size of R_i and R_o for the non-inverting amplifier?
- why the voltage follower is a special case of the non-inverting amplifier?
- the A_v of the voltage follower?
- the phase difference between the input and output of the voltage follower?
- how a voltage follower is used to improve a poor voltage source?
- other common names for the voltage follower circuit?

Tutorial questions

1   What type of feedback is used to stabilize circuits?
2   If the feedback causes the output to be reduced in size, what type is it?
3   Is it possible to feed back anything other than voltage? If so, what?
4   What is the condition which prevents us from applying the two op-amp rules?
5   Draw an inverting amplifier circuit, choosing typical resistance values for an inverting voltage gain of 80. What is the input resistance of the circuit you have drawn?
6   An inverting amplifier circuit has an $R_i = 100$ kΩ and $A_v = -10$. What size of feedback resistance is being used?
7   The gain of an inverting amplifier is set at -20 by using 1 kΩ and 20 kΩ resistances. The circuit is driven from a 1 V voltage source whose output resistance is 1 5 kΩ. What is the V_{out} from the op-amp circuit?
8   A certain inverting amplifier has $R_i = 7$ 5 kΩ and V_{in} comes from an ideal 9 V supply. What is the size of the feedback current and V_{out} if the feedback resistance is set at 3 5 kΩ, 5 kΩ and 7.5 kΩ?
9   What is an op-amp non-inverting amplifier with $A_v = 1$ called?
10 What are the approximate input and output resistances of a non-inverting amplifier?

11 Draw a non-inverter with appropriate resistances to give an A_v of 8.

12 What are two other names for the voltage follower circuit?

13 How and why would a buffer circuit improve voltage transfer between a source and a load? Give an example, using different values from that in the text

14 What is the feedback fraction of a buffer circuit?

15 The feedback resistor of a non-inverter circuit has been chosen as 15 kΩ and the circuit A_v is 10. What size is the other resistor in the circuit?

16 If V_{in} = 1 V in the circuit of question 15, what size is the current flow in each resistor?

17 If V_{in} = 2 V in the circuit of question 15, what size is the current flow in each resistor if the feedback resistor value is reduced to 5 kΩ?

18 What effect on the action of a non-inverting circuit would there be if a series resistor were introduced in the input wire?

19 Design a non-inverting circuit to give a gain of 500 for a V_{in} of 10 mV, while limiting the current flow from the output of the op-amp to 10 mA.

20 What would happen if a non-inverting circuit with A_v = 50 developed the following faults: the feedback resistor became short circuited, the resistor to earth became open-circuited and the feedback resistor became open-circuited?

Further reading

Bogart T.F. (1986). *Electronic devices and circuits*, section 13 1. Merrill, Columbus, Ohio

Horowitz P & Hill W (1987) *The art of electronics*, sections 3 01–6, 3 24–5 Cambridge University Press, Cambridge, England

CHAPTER 4

Op-amp circuits in regular use

Aims and objectives

The aim of this chapter is to build up your knowledge of standard op-amp circuits in regular use It will also give you the chance to become more familiar with the ideal op-amp analysis used in Chapter 3 and give you the confidence to analyse non-standard op-amp circuits you may see used in other literature The key to being a good circuit designer is to have at your fingertips a library of circuits which will perform specific tasks and to know how each circuit works.

When you have studied this chapter you will be able to design circuits which will·

- *detect current flow*
- *sum a number of different input currents*
- *sum a number of different input voltages*
- *perform integration and differentiation*
- *detect and amplify an unearthed potential difference*

Other than the differentiator and the integrator, the circuits in this chapter are linear circuits This means that we can apply the two simple rules that were used in Chapter 3.

We will be using the same analysis technique as that used in Chapter 3, but we will be more interested in an expression for the output voltage V_{out} rather than the voltage gain A_v. This is partly because we will not always be driving the circuit with an input voltage. R_{in} and R_{out}, however, will still be needed.

The current sensor, the current adder and the voltage adder are all variations on the standard inverting circuit and, as you will see, require very little extra in the way of analysis.

4.1 The current sensor

Other names for the current sensor are 'transresistance amplifier' and 'current-to-voltage converter' The last name is the most descriptive. This circuit takes as input a current which is converted into an output voltage The conversion is linear so that

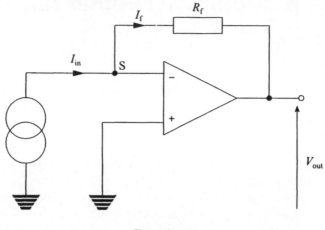

Figure 4.1

if the input current doubles, so does the output voltage. Look at Figure 4 1 for a moment. You will see that this circuit is almost the same as the inverting amplifier: the only difference is that the first resistance (R_1) has been removed from the input connection of the inverter. Since we now have only one resistance in the circuit, which is the feedback resistance, we call it R_f. It gives us an output voltage V_{out} equal to the input current I_{in}, multiplied by the feedback resistance R_f

$$V_{out} = -I_{in} \times R_f$$

Output voltage
Let us see how we obtain the output voltage of a current sensor circuit. With reference to the circuit diagram in Figure 4.1, using rule 1 we can say

point S is a virtual earth

Using rule 2

$$I_{in} = I_f$$

but

$$I_f = (0 - V_{out})/R_f$$

$$= -V_{out}/R_f$$

Therefore

$$I_{in} = -V_{out}/R_f$$

or

$$V_{out} = -I_{in} \times R_f$$

Input resistance

As with the inverter, we need also to investigate the input resistance value R_{in} of the current sensor This is easy! Input resistance is always defined as $R_{in} = V_{in}/I_{in}$. For this circuit, however, the voltage at the input is a virtual earth 0 V! Therefore

$$R_{in} = 0/I_{in}$$

$$R_{in} = 0\,\Omega$$

There are three points to note about this circuit

1   The minus sign tells you that an input current entering the circuit will result in a negative output voltage. This also means that if a current flows in the opposite direction (i.e. *out* of the circuit), the resulting voltage will be positive.
2   The size of the output voltage will be governed by the choice of feedback resistance R_f
3   The circuit offers no resistance whatsoever to any circuit connected to its input.

EXAMPLE 4.1

We want to watch and measure, on a cathode ray oscilloscope (CRO), the current variations in a particular circuit of $100\,\mu A$ to 1 mA.

SOLUTION

Unfortunately, a CRO cannot accept current inputs· it has only voltage inputs We must therefore convert the current variations into voltage variations How? We use the current sensor (Figure 4 2) The minimum $I_{in} = 100\ \mu A$ Therefore

$$V_{out(min)} = -100\ \mu A \times 10\ k\Omega$$

$$= -1\ V$$

The maximum $I_{in} = 1$ mA Therefore

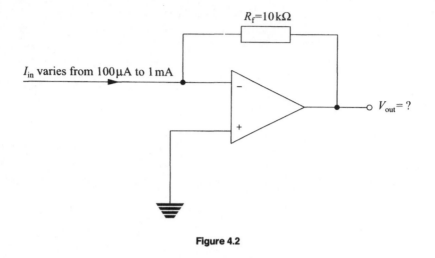

Figure 4.2

$$V_{out(max)} = -1 \text{ mA} \times 10 \text{ k}\Omega$$

$$= -10 \text{ V}$$

The V_{out} can now be monitored safely on the CRO.

If the current in this example were reduced to one-tenth of its value, we could easily amplify it back up to the required level for the CRO by increasing R_f to 100 kΩ

Self-assessment questions

SAQ4.1 What does the minus sign mean?

SAQ4.2 What is the input resistance of the current sensor circuit?

SAQ4.3 If a current sensor circuit has R_f = 2 kΩ and I_{in} = 2 5 mA, what size is V_{out}? (Look at your answer It is negative Yes? Should it be? Does it matter which way the input current is flowing?)

SAQ4.4 A current sensor has R_f = 0 1 kΩ and an input current flowing away from the op-amp I_{in} = 5 mA What is V_{out}?

Output resistance
Remember that the current sensor is merely another form of the standard inverting amplifier. Therefore, the R_{out} value is just the same! Very low We can approximate this to

$$R_{out} = 0 \, \Omega$$

as before

4.2 The current adder

The input current of the current sensor circuit may often be made up of a number of currents from the outputs of different circuits. If it is, hey presto! We have the current adder shown in Figure 4.3.

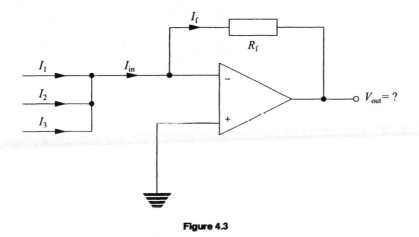

Figure 4.3

Output voltage
The current adder circuit is almost the same as the current sensor circuit, the only difference being that in the former the input current forced around the feedback loop is the sum of the incoming external currents We have therefore

$$I_{in} = I_f$$

But

$$I_{in} = I_1 + I_2 + I_3$$

and (as before)

$$I_f = -V_{out}/R_f$$

Therefore

$$-V_{out}/R_f = I_1 + I_2 + I_3$$

giving

$$V_{out} = -(I_1 + I_2 + I_3) \times R_f$$

Input and output resistances

We can also say for completeness that the input and output resistances for the current adder circuit are the same as those for the current sensor circuit, i e

$$R_{in} = 0\,\Omega$$

$$R_{out} = 0\,\Omega$$

It must be said, however, that the current adder is not as useful as the next circuit in our collection. the voltage adder

4.3 The voltage adder

There are many times in electronics when we want to add two or more voltages together. This task is easily performed by the voltage adder circuit, shown in Figure 4 4 Notice that this also is only a variation of our now familiar inverting amplifier circuit.

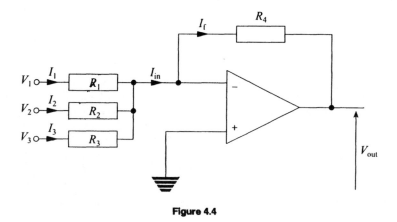

Figure 4.4

Output voltage

V_{out} is equal to the sum of the input voltages We have a sum of input voltages but we can make it add different amounts of the input voltages. To understand this more fully, let us first see how the output voltage expression is derived.

In Figure 4.4 we have three currents flowing into the circuit, just as we did with the current adder circuit Therefore, we can go straight to the current adder expression

$$V_{out} = -(I_1 + I_2 + I_3)/R_f$$

But of course we can rewrite each input current in terms of an input voltage Remembering that we have a virtual earth at the inverting input, we get

$$I_1 = V_1/R_1$$

$$I_2 = V_2/R_2$$

$$I_3 = V_3/R_3$$

Substituting these expressions into the output voltage expression, we get

$$V_{out} = -(V_1/R_1 + V_2/R_2 + V_3/R_3) \times R_f$$

Multiplying out gives

$$V_{out} = -(V_1 \times R_f/R_1 + V_2 \times R_f/R_2 + V_3 \times R_f/R_3)$$

This tells us that each input voltage is added to the other input voltages to give a summed output voltage, but note that we can (if we want to) add different weights to each input by varying its associated resistance For example, if we wanted to add three times more of V_1 than of V_2, we would merely make R_1 three times smaller than R_2 Note also that R_f is common to them all!

EXAMPLE 4.2

Let us say that we have two input voltages $V_1 = 2\,V$ and $V_2 = 3\,V$, and $R_f = 5\ k\Omega$ Now, to make V_1 twice as important in the summation as V_2, we could choose $R_1 = 2\ 5\ k\Omega$ and $R_2 = 5\ k\Omega$ Find V_{out}

SOLUTION

$$V_{out} = -[2 \times (5/2.5) + 3 \times (5/5)]$$

$$= -[4 + 3]$$

$$= -7\ V$$

Note the minus sign

Perhaps you only want a circuit to add voltages together, *without* weighting the inputs differently. If so, all you do is make all the input resistors the same size! Then, since

$$R_1 = R_2 + R_3 = R$$

it follows that

$$V_{out} = -(V_1 + V_2 + V_3) \times R_f/R$$

EXAMPLE 4.3

We want to add together the outputs from three circuits $V_1 = 0.2\,V$, $V_2 = 0.4\,V$, $V_3 = 0\,5\,V$ To do this we must choose the resistance values. Let us choose 1 kΩ for all of them Find V_{out}

SOLUTION

$$V_{out} = -(V_1 + V_2 + V_3) \times R_f/R$$

Therefore

$$V_{out} = -(0.2 + 0.4 + 0.5) \times 1/1$$
$$= -1.1\ V$$

EXAMPLE 4.4

We now might decide that we want to amplify the final output by, say, 10.

SOLUTION

All we do is to increase R_f from 1 kΩ to 10 kΩ! Then

$$V_{out} = -11\ V$$

Output resistance

Knowing that the voltage adder circuit is just the inverter with extra inputs, it should mean that you know its output resistance straight away

$$R_{out} = 0\,\Omega$$

Input resistance

Since the voltage adder circuit is just the inverter with more than one input, you can probably guess its input resistance· the R_{in} value for each input is dependent on the size of its input resistor.

Self-assessment questions

SAQ4.5 What is the R_{in} for each input in Example 4.3?

SAQ4.6 Design an op-amp voltage adder to add 0.5 V, 0 03 V, 4 1 V and 2 04 V with equal weighting and equal input resistance of 10 kΩ at each of its inputs.

Using a.c. signals with the voltage adder
One of the advantages of an op-amp circuit is its ability to process a c. or d c. signals Try the question below

Self-assessment question

SAQ4.7 We want to design an audio mixing circuit which will add together two audio signals at varying volume levels We therefore use a two-input voltage adder circuit, but we use variable resistances for R_1, R_2 and R_f It is best to draw the circuit. Now consider the effect of changing each of the resistance values, one at a time What will happen to the sound output?

The audio mixer circuit described in SAQ4 7 works fine, but one of its problems is that the R_{in} at each input varies as we alter the volume controls at each input. The simple and effective way to fix this is to insert an op-amp buffer circuit (remember, unity gain voltage followers) between the input voltage sources and the voltage adder circuit. If you are not sure about this, look back to Chapter 3 at the section on the application of the voltage follower as a buffer circuit

The final three circuits in this chapter by no means complete the subject on op-amps, but do complete your introductory portfolio of standard op-amp circuits. Here we see one or two more analysis concepts creeping in, e.g. the law of superposition which is used in the derivation of V_{out} for the difference amplifier. We also see capacitances used for the first time in op-amps

4.4 The op-amp differentiator

The word 'differentiator' will probably make you think of mathematics (calculus to be specific) and this is exactly what the circuit does. it performs a mathematical operation on the input voltage The operation it performs, as you might guess, is differentiation Those of you who have studied some calculus will know that this means finding the slope or gradient of a function at a point. This differentiator circuit will differentiate the input voltage V_{in} with respect to time. In other words, it will tell you the rate of change of V_{in} with respect to time, i.e. dV/dt or, more specifically, dV_{in}/dt.

The circuit is known as a 'non-linear' circuit. This basically means that a linear increase in input voltage (or current) does not necessarily give us a linear increase in output. This means that the output waveform can take on a different shape from that of the input waveform It was originally used as a mathematical tool to help engineers simulate real applications by using mathematical models. The circuit formed part of a large network of analog circuits and was called an 'analog computer'. Nowadays, however, differentiators (and integrators) are readily used as signal processing circuits and therefore still have many uses

Uses

If we differentiate a sine wave, we obtain a cosine wave (which is a phase-shifted sine wave). If we differentiate a triangle wave, we obtain a square wave. If we differentiate a square wave, we obtain a spike waveform, with spikes alternately positive and negative. Why would we want to do these things? This will become more apparent when you start designing! Let us look, therefore, at the differentiating circuit in Figure 4.5.

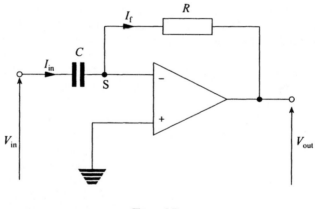

Figure 4.5

Analysis

To analyse the differentiator circuit using our op-amp rules, we must first note that we have a capacitance in the circuit Capacitance theory tells us that the capacitor current I_c is proportional to the rate of change of voltage across its plates dV/dt and to the size of the capacitance C in farads. This yields the equation

$$I_c = C\,dV/dt$$

In other words, if we change the potential across a capacitor very quickly, we will obtain a current larger than if we make the same change more slowly. Using this piece of information, we can start our analysis

Output voltage

Application of rule 1 shows that point S is a virtual earth and, using rule 2, we get

$$I_f = I_{in}$$

But I_{in} is the capacitor current and the voltage across the capacitor is always ($V_{in} -$ 0) Therefore, we can say

$$I_{in} = C\,d(V_{in} - 0)/dt$$

Therefore

$$I_{in} = C\,dV_{in}/dt$$

Also

$$I_f = (0 - V_{out})/R$$

Therefore

$$I_f = -V_{out}/R$$

Substitution into $I_f = I_{in}$ gives

$$-V_{out}/R = C\,dV_{in}/dt$$

Therefore, finally

$$V_{out} = -CR\,dV_{in}/dt$$

This means that an output voltage will be produced which is a differentiated version of the input Note the negative sign, which tells you that the circuit is still an inverter. The size of V_{out} is dependent on the sizes of C and R. CR is known generally as the 'time constant' and, in this circuit, CR is also known as the 'gain factor'.

EXAMPLE 4.5

If $C = 1\ \mu F$, $R = 1\ M\Omega$ and V_{in} is increasing at a rate of 1 volt per second ($dV_{in}/dt = 1\ V\ s^{-1}$), find V_{out}

SOLUTION
We would have

$$V_{out} = -(1 \times 10^{-6})(1 \times 10^{6})(+1)$$
$$= -1 \text{ V}$$

That is, the output is the slope or gradient of the input at any instant, multiplied by a fixed constant CR. If the slope is constant, the output is constant. If the slope varies, the output varies. Note the minus sign. This is still an inverting amplifier; therefore, a positive slope at the input gives a negative output.

Self-assessment question

SAQ4.8 Sketch a waveform diagram showing both input and output of a differentiator driven with a sine wave. (Do not worry for the moment about the amplitude.)

Let us now look at an example involving amplitudes

EXAMPLE 4.6
Assume we have the same C and R values as in Example 4 5. Now assume an input sine wave of 1 mV_{rms} at 10 Hz. What does the output look like?

SOLUTION
First, turn V_{in} into a peak value

$$V_{peak} = V_{rms} \times 1.414$$
$$= 1 \times 1.414$$
$$= 1.4 \text{ mV}_{peak}$$

So

$$V_{in} = (1.4 \times 10^{-3})\sin \omega t$$

Now, using mathematics to differentiate a sine function

$$dV_{in}/dt = (1.4 \times 10^{-3}) \times \omega \times \cos \omega t$$

Remember, $\omega = 2\pi f$ Therefore

$$\omega = 2 \times 3.14 \times 10$$
$$= 62.8 \text{ rad s}^{-1}$$

Therefore

$$dV_{in}/dt = (1.4 \times 10^{-3}) \times 62.8\cos\omega t$$
$$= (87.9 \times 10^{-3})\cos\omega t$$

Now the circuit performs this differentiation, but it also multiplies by CR and inverts, giving

$$CR = (1 \times 10^{-6}) \times (1 \times 10^{6})$$
$$= 1$$

Therefore

$$V_{out} = -(1) \times (87.9 \times 10^{-3})\cos\omega t$$
$$\approx -(88 \times 10^{-3})\cos\omega t$$

The result of this is that the output voltage amplitude is 88 mV, and it is a negative cosine wave

Self-assessment question

SAQ4.9 That was the hard bit. Now you sketch the input and output waveforms (superimpose them to show the phase relationship). Drawing to scale is not necessary, but labelling everything (including V_{peak} values) is!

Interestingly enough, there is a quicker and easier way of looking at this problem. You can consider the differentiator to be a simple inverter circuit but, instead of using $-R_2/R_1$ as the gain, you use $-R/Z$ (where R is the feedback and Z is the reactance of the input capacitor at the specified frequency).

EXAMPLE 4.7
Solve Example 4.6 using the equation $A_v = -R/Z$.

SOLUTION

$$Z = 1/\omega C$$
$$= 1/(2 \times 3.14 \times 10 \times 1 \times 10^{-6})$$
$$= 15.9 \text{ k}\Omega$$
$$R = 1 \text{ M}\Omega = 1000 \text{ k}\Omega$$

Therefore

$$A_v = -R/Z$$

$$= -1000/15.9$$

$$= -62.8$$

Now an input sine wave will be amplified by -62.8.
If the input peak voltage is $1.4\,\text{mV}$, we have

$$V_{out(peak)} = 1.4 \text{ mV} \times 62.8$$

$$= 88 \text{ mV}$$

but you must also realize that the circuit differentiates and inverts, giving

$$V_{out} = -(88 \times 10^{-3})\cos \omega t$$

as before!

Now you try an example.

Self-assessment questions

SAQ4.10 The input to an op-amp differentiator is $10 \sin \omega t$ with $f = 1000$ Hz For $C = 0.01\ \mu\text{F}$ and $R = 20$ kΩ, draw V_{in} and V_{out} superimposed on a waveform diagram and label their amplitudes.

SAQ4.11 Look at the previous question and consider what would happen to V_{out} if the input remained the same size but increased in frequency. Work out V_{out} for f values of 2 kHz, 10 kHz, 20 kHz and 50 kHz.

SAQ4.12 Using the answers from the previous question, imagine what might happen if we try to differentiate an input signal of low frequency with a very small amount of high frequency noise superimposed on the signal. What might be the result? The result is one reason why differentiators are avoided if possible! You might find this difficult. Look at the answer after giving it some thought

A close relative of the op-amp differentiator is the op-amp integrator.

4.5 The op-amp integrator

The introduction to the op-amp differentiator could also be read for the integrator
Since integration is the inverse of differentiation, we can use it to obtain triangle
waves from square waves, sine waves from cosine waves and square waves from
spikes Hence it is another non-linear signal processing circuit in common use Its
action produces an output voltage which represents the area under the curve of the
input signal, from one point in time to another (This is one way of calculating
definite integrals in mathematics.) Note that we can only integrate with respect to
time We again have a circuit containing a capacitor. Look at the integrator in
Figure 4 6.

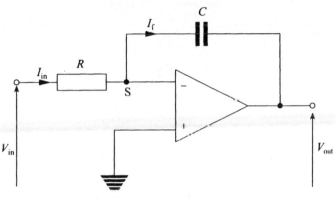

Figure 4.6

Output voltage
Using what we already know, using rule 1 we can say that point S is a virtual earth
and, using rule 2, we get

$$I_{in} = I_f$$

But I_f is the capacitor current and the voltage across the capacitor is always $(0 -
V_{out})$ Therefore, we can say

$$I_f = C \ d(0 - V_{out})/dt$$

Therefore

$$I_f = -C \ dV_{out}/dt$$

Also

$$I_{in} = (V_{in} - 0)/R$$

Therefore

$$I_{in} = V_{in}/R$$

Substitution into $I_{in} = I_f$ gives

$$V_{in}/R = -C \, dV_{out}/dt$$

Therefore

$$dV_{out}/dt = -V_{in}/CR$$

Integrating to obtain V_{out}, we get

$$V_{out} = -1/CR \int_{t_1}^{t_2} V_{in} \, dt$$

This means that an output voltage will be produced which is an integrated version of the input voltage over the time interval from t_1 to t_2 seconds Note also that it is an inverter (like the differentiator) and that the size of V_{out} is dependent on the values of C and R. In this circuit, $1/CR$ is called the 'gain factor'. Notice here that the input is integrated *then* multiplied by $-1/CR$

EXAMPLE 4.8
Find V_{out} if $CR = 1$ and $V_{in} = +1$ V.

SOLUTION
As $CR = 1$, it follows that $1/CR = 1$. A constant $+1$ V input would result in an output (V_{out}) falling with a negative slope of 1 V s^{-1} from zero to (theoretically) infinity.

Self-assessment question

SAQ4.13 Draw the appropriate V_{in} and V_{out} waveforms for Example 4 8 and choose appropriate C and R values.

The integrator circuit is often used to generate a ramp waveform for timing purposes. You can see from Example 4.8 that constant or d.c. inputs cause V_{out} to reach very large positive or negative values. In reality, these large values are unattainable and, in fact, unwanted. The circuit, therefore, usually has as an input a periodically changing d.c. level (a square wave with zero d.c. offset) which produces a triangle output swinging between two output voltage limits (e.g. +10 V and −10 V), or the capacitor is periodically discharged (short-circuited) through a mechanical or solid-state switch. The circuit integrates, which obviously means that at any instant it will give an output equal to the area swept out by the input voltage (i e. the area under the voltage/time curve) from time t_1 to time t_2.

Self-assessment questions

SAQ4.14 An integrator with $C = 1\,\mu F$ and $R = 100\,k\Omega$ is used. The input at time zero rises instantly from 0 V to 1 V and stays constant for 1 ms. After 1 ms it drops instantly back to 0 V. Determine the V_{out} waveform and draw V_{in} and V_{out} on a waveform diagram. (Hint: find CR and hence $1/CR$, then remember that the circuit integrates V_{in} from zero to 1 ms (area $= V_{in} \times t$). Finally, remember the inverting action!)

SAQ4.15 Use the same circuit as that in the previous question and find the output for an input $5\,V_{peak-peak}$ square wave (with no d.c. offset) at a frequency $f = 500\,Hz$

SAQ4.16 What do you think would happen to the amplitude of V_{out} in the previous question if the frequency were increased to 1000 Hz?

The last single op-amp circuit we are going to look at is the difference amplifier.

4.6 The difference amplifier

The difference amplifier is also called a 'differential amplifier' or 'balanced input amplifier'. It is another voltage amplifier, but it differs from any other we have looked at. Why? Well, so far we have only considered voltage amplifiers with input voltages measured with respect to earth Look, for example, at the voltage adder circuit. V_1 is an input voltage applied with respect to earth, i.e. the input is earth-referenced

Input voltages not earth-referenced
Quite often we have input voltages which are not earth-referenced A voltage is a potential difference measured between two points in a circuit. If one of the points is not at 0 V, then the reading is not earth-referenced. Look at the potential divider circuit in Figure 4.7 You can see that V_{out} is measured across a resistance and is not earth-referenced. How, therefore, can we use this as an input to a voltage amplifier? Until now we could not: that is, until we met the difference amplifier. The difference amplifier looks slightly more complicated than the previous circuits we have considered. Look at the difference amplifier shown in Figure 4 8.

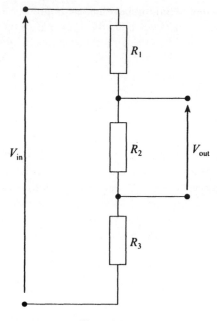

Figure 4.7

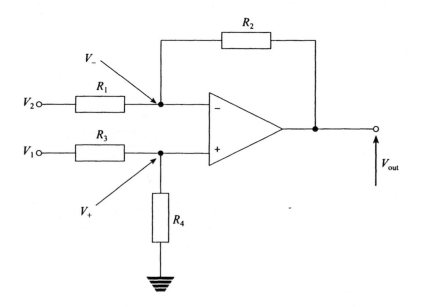

Figure 4.8

Three points are worth noting here·

1. that the input voltage to the circuit is between two input connections, and is the difference between the voltages V_1 and V_2;
2. that, to make the analysis a little easier to follow, the voltages at each of the op-amp inputs have been labelled V_- and V_+;
3. that we are interested in an expression which relates V_{out} to V_{in}, in terms of the resistance values.

Using the law of superposition for analysis

One useful method of analysis for the difference amplifier circuit is application of the law of superposition This law allows us to find the total output voltage by considering the separate effects of each input voltage

We use it by applying the following argument To find the total output voltage we must consider how the output is related to changes on each of the inputs. We can do this by considering each input separately (we merely short-circuit to ground any voltage inputs not being considered) and then adding all the effects together at the end By adding all the separate effects together, we obtain an expression which relates the output to all the inputs In this case only two inputs are considered

Output voltage and voltage gain

Using superposition, consider V_1 first Then short-circuit other voltage sources to ground (sh–cct V_2). It is then usually helpful to redraw the cct, as in Figure 4.9 Note that the op-amp is drawn the other way up

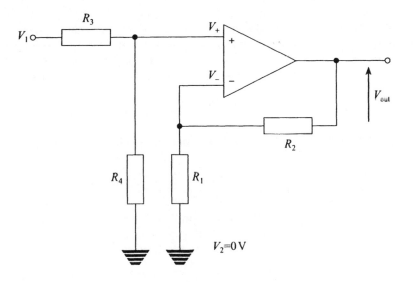

Figure 4.9

If you ignore V_1, R_3 and R_4 for a moment and just look at V_+, you can see that V_+ is now an input to a standard non-inverting amplifier with a gain of $1 + R_2/R_1$. We could say

$$V_{out}/V_+ = 1 + R_2/R_1$$

$$= (R_1 + R_2)/R_1$$

(Remember this from non-inverting op-amp analysis?) But V_+ is not the input to this circuit: the input is V_1. V_+ is just the output of the V_1, R_3, R_4 voltage divider, i.e. by voltage divider action

$$V_+ = V_1 \times R_4/(R_3 + R_4)$$

Therefore, the right-hand side of this expression is the new input voltage to the non-inverter. This gives

$$V_{out}/[V_1 \times R_4/(R_3 + R_4)] = (R_1 + R_2)/R_1$$

Therefore

$$V_{out} = [V_1 \times R_4/(R_3 + R_4)][(R_1 + R_2)/R_1]$$

Now consider V_2. Short-circuit V_1 and redraw the circuit as shown in Figure 4.10 Now, the sh-cct of V_1 gives us a parallel R_3 and R_4 combination. If we ignore R_3 and R_4, what we have left is a simple inverting amplifier, with V_2 as input and R_2/R_1 as the gain. Can we ignore the fact that R_3 and R_4 are also present in the circuit? Yes, because the second ideal op-amp rule tells us that the op-amp draws no current. Therefore, R_3 and R_4 carry zero current and have zero voltage drop across them. This means that the op-amp non-inverting input is still at zero voltage and the inverting circuit action continues as normal. Therefore, we can say

$$V_{out} = -V_2 \times R_2/R_1$$

Adding together these last two equations for V_{out} gives the following expression for the total V_{out}

$$V_{out} = [V_1 \times R_4/(R_3 + R_4)][(R_1 + R_2)/R_1] - [V_2 \times R_2/R_1]$$

This, however, is not an easy or manageable expression. The circuit performs better anyway if resistance values at each input are symmetrical. Therefore, we normally choose to make $R_1 = R_3$ and $R_2 = R_4$, and so the expression for V_{out} cancels down to

$$V_{out} = (V_1 - V_2) \times R_2/R_1$$

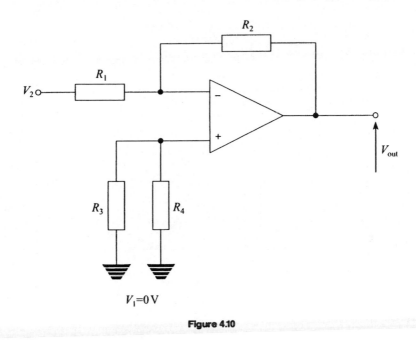

Figure 4.10

Hence, the difference in input voltage $V_{\text{in}} = (V_1 - V_2)$ is amplified by R_2/R_1 to give V_{out}.

We could express the difference gain $A_{\text{v(diff)}}$ as

$$A_{\text{v(diff)}} = V_{\text{out}}/V_{\text{in}}$$

$$= R_2/R_1$$

(Note that there is no inversion)

Finally, we can make all the resistances equal to give a unity gain difference amplifier

$$A_{\text{v(diff)}} = 1$$

$$V_{\text{out}} = V_1 - V_2$$

The full expression for the op-amp difference amplifier output is too long for you ever to bother to memorize it but, as you can see, by choosing certain resistance values the expression becomes manageable. Notice that the difference amplifier gain can easily be adjusted to our requirements by using certain R values. For instance, if we choose $R_2 = 10 \text{ k}\Omega$ and $R_1 = 1 \text{ k}\Omega$, we have an $A_{\text{v(diff)}}$ of 10.

Common mode voltages

If both inputs are at the same voltage, the input voltage is called a 'common mode input' (or 'common mode signal') and, since the circuit will deal only with

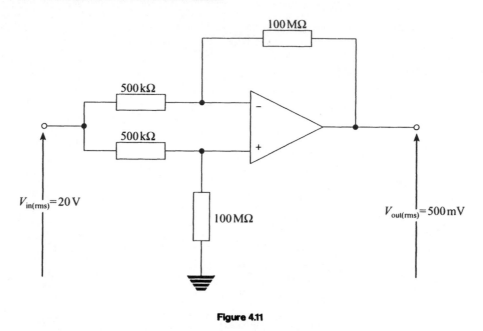

Figure 4.11

difference voltages at the input, it is obvious that if both inputs are at the same voltage, there is no difference and therefore no output. If only things were perfect! A real difference amplifier circuit responds very slightly to voltages that appear on both inputs at once, known as 'common mode signals'. This may be explained with reference to Figure 4.11. If both inputs are connected together so that they become a common input to the circuit, no difference voltage can ever be produced (as the inputs are shorted together), but both together they could undergo large changes in voltage (common mode). When this happens, the output produces a very small version of this common mode signal.

For example, we might find that a common mode signal of 20 V amplitude at the input produced a 0.5 V version of this signal at the output. (Notice that this output is of the order of 40 times smaller than the input!) This is okay as we want any common mode input signals to be as small as possible (ideally zero) at the output. Common mode signals are not wanted, but they sometimes appear anyway and a difference amplifier is a good way of reducing their effect. We can measure the gain that the circuit has on these unwanted common mode signals by deliberately applying a common mode input ($V_{in(cm)}$) and then measuring the resulting common mode output ($V_{out(cm)}$). The common mode gain is called $A_{v(cm)}$), where

$$A_{v(cm)} = V_{out(cm)}/V_{in(cm)}$$

Common mode rejection ratio
A measure of the quality of a difference amplifier is how good it is at rejecting common mode signals and processing only difference signals This quality is

defined as the 'common mode rejection ratio' (CMRR), where

$$\text{CMRR} = \text{difference gain/common mode gain}$$

i e

$$\text{CMRR} = A_{v(\text{diff})}/A_{v(\text{cm})}$$

Hopefully this is a large number It is sometimes expressed in dB

$$\text{CMRR} = 20 \log_{10}(A_{v(\text{diff})}/A_{v(\text{cm})})$$

Self-assessment questions

SAQ4.17 The CMRR for a certain circuit is quoted as 90 dB How much bigger is the difference gain over the common mode gain?

SAQ4.18 Look at the difference amplifier circuit diagram in Figure 4.11 and work out the difference gain, the common mode gain and the CMRR (in dB)

The reason for considering common mode signals

Considering common mode signals at all might have confused you up to now. The reason is probably best tackled by giving an example of a typical application of a difference amplifier circuit.

CASE STUDY

A signal from a microphone or guitar on stage must be carried to the mixing desk at the back of a large hall. It is desirable that the signal does not pick up lots of interference (or noise) on its way! On its travels, however, it passes very close to many sources of interference, e.g. 50 Hz mains hum from power supply cables and random electromagnetic interference from stagelights being switched rapidly on and off If the microphone signal is earth-referenced and drives a standard inverting or non-inverting amplifier, the signal and the interference are amplified equally! See Figure 4.12 This is not a desirable situation, since the output contains lots of unwanted interference as well as the microphone signal. This is where the difference amplifier comes into its own.

First of all we unearth the microphone to provide us with a difference signal which is not earth-referenced. Then we use a difference amplifier instead of the non-inverter What happens now is that the unwanted interference appears on both the inputs simultaneously. We must now remember that if a signal appears simultaneously on both inputs, it will be seen as a common mode signal and will not be amplified. This is illustrated in Figure 4 13. Note that the microphone signal is a difference signal (i e it is not the same at both inputs to the amplifier)

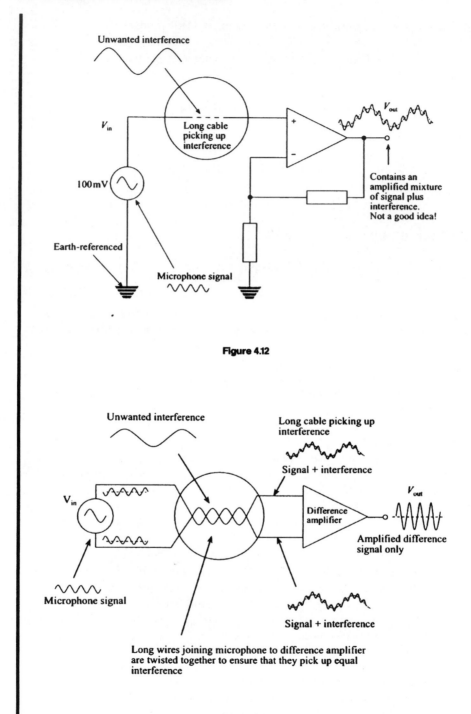

Figure 4.12

Figure 4.13

and is therefore amplified by the difference voltage gain of the circuit $A_{v(diff)}$ $A_{v(diff)}$ can be made quite large depending on the resistor ratios chosen.

One final point must be made here about the interference picked up by the long cable We usually try to reduce the amount of interference picked up by housing the insulation-covered, twisted cable (called a 'twisted pair') *inside* a metallic sheath connected at one point in the circuit to earth. In this way the common mode interference picked up is minimized by shorting some of it to the earth! This type of connection to the difference amplifier is called a 'balanced line'.

This chapter has given you an introduction to some of the basic op-amp circuits and their appropriate output voltage equations

Checklist

Do you know:
- how to draw the circuit diagrams of the current sensor, current and voltage adder?
- the R_{in} and R_{out} of the current adder?
- the R_{in} and R_{out} of the voltage adder?
- for which condition the voltage adder output is given by $V_{out} = -(V_1 + V_2 + \ldots + V_n)$?
- the output voltage expressions for the current and voltage adder circuits?
- whether or not all these circuits are called 'linear circuits'?
- whether or not all these circuits are called 'inverting circuits'?
- the difference in the diagrams of the differentiator and integrator circuits?
- what the gain factor for each of the differentiator and integrator circuits is?
- what happens to V_{out} of the differentiator as the input frequency increases?
- what the circuits will differentiate and integrate with respect to?
- how to generate a triangle waveform from a square wave?
- what a non-earth-referenced input voltage is?
- what CMRR means?
- how to measure the CMRR of a circuit?
- how to reduce interference in a long cable carrying a small signal?
- what 'twisted pair' and 'balanced line' mean?

Tutorial questions

1 What size and direction is I_{in} if a current sensor has $R_f = 10 \, k\Omega$ and $V_{out} = 10 \, V$?
2 What size and direction is I_{in} if a current sensor has $R_f = 6.8 \, k\Omega$ and $V_{out} = -9 \, V$?
3 What are two other names for the current sensor circuit?
4 Does a current sensor have a virtual earth?

5   Could we have good voltage transfer into a current sensor circuit?

6   Design a current sensor to give V_{out} in the range 1–10 V for an input current range of 10–100 μA.

7   Design a voltage adder to add $V_1 + 2V_2$.

8   What is the input resistance of each of the inputs in your answer to question 7?

9   Design a voltage adder to add $10V_x + 4V_y$. Specify the two input resistance values.

10   Draw an op-amp differentiating circuit with $C = 2.5 \mu$F and $R = 1 k\Omega$. If V_{in} is a 1 V_{rms} sine wave at 10 Hz with zero d.c. offset, calculate the amplitude of V_{out}. Draw V_{in} and V_{out} on a waveform diagram.

11   For the circuit in question 10, calculate the new V_{out} values if the input frequency were changed to 50 Hz, 150 Hz, 5 Hz and 1 kHz.

12   An integrator with $C = 0.001 \mu$F and $R = 10 M\Omega$ is used. The input at time zero rises instantly from 0 V to 5 mV and stays constant for 2 ms. After 2 ms it rises instantly to 15 mV and stays constant for a further 2 ms. It then drops instantly back to 0 V again. Determine the V_{out} waveform and draw V_{in} and V_{out} on a waveform diagram.

13   Draw the general result of differentiating a sine wave, a square wave, a triangle wave and a sawtooth wave.

14   Draw the general result of integrating the following waves, assuming that they all have a zero d.c. offset: a sine wave, a square wave, a triangle wave and a sawtooth wave.

15   The CMRR for a certain circuit is quoted as 65 dB. How much larger is $A_{v(diff)}$ than $A_{v(cm)}$?

16   If CMRR = 50 dB and $A_{v(cm)} = 10^{-8}$, what is the value of $A_{v(diff)}$?

17   If $V_{in(cm)} = 5$ V produces a $V_{out(cm)}$ of 100 mV and all the difference amplifier resistors are equal, what is the CMRR value in dB?

18   If a difference amplifier is used to reduce interference in a long cable, would we prefer a very high or a very low CMRR value?

19   Does an op-amp difference amplifier circuit have a virtual earth?

20   If $A_{v(cm)} = 0.01$ and $A_{v(diff)} = 10$, what would be the V_{out} value for an input of $V_1 = 11.3$ V and $V_2 = 10\,9$ V? (Hint: you need to find values of $V_{in(cm)}$ and $V_{in(diff)}$.)

Further reading

Bogart T.F (1986) *Electronic devices and circuits* Merrill, Columbus, Ohio

Floyd T L (1988). *Electronic devices*, 2nd edn Merrill, Columbus, Ohio

Floyd T L. (1993). *Principles of electric circuits*, 4th edn, pp 547–64 Merrill, Columbus, Ohio.

Horowitz P. & Hill W (1987) *The art of electronics*, sections 3 09 and 3 18–19 Cambridge University Press, Cambridge, England.

CHAPTER 5

Op-amp limitations

Aims and objectives

This chapter will give you more of an insight into the differences between ideal and real op-amp circuits. It will demonstrate the particular requirements, specifications and limitations of real op-amps used in standard circuits. You will also see how, by the judicial use of certain component values, you can still treat op-amps as being ideal for many applications.

After working through the chapter you will be conversant with the following op amp definitions:

- *input bias current*
- *input offset current*
- *input offset voltage*
- *open loop gain*
- *bandwidth*
- *gain bandwidth product*
- *slew rate*
- *output saturation levels*

You will also know how to modify your ideal op-amp circuits (covered in Chapters 3 and 4) where necessary, to account for these factors. We will look at the inverting and non-inverting op-amp amplifier circuits.

5.1 Real op-amps

In previous chapters you have been shown some simple op-amp circuits and you should now have some understanding of how they can be analysed using simple ideal op-amp rules Now we are going to look at some of the things that happen when you use a *real* op-amp in a circuit. You may, of course, already know about some of these things if you have done any practical work involving op-amps. Throughout this chapter, wherever a typical value for a real op-amp is quoted, it will refer to the 741 mentioned in earlier chapters. So we now have a starting point for looking at some of the ways in which a real op-amp circuit differs from our ideal concept. Although a knowledge of these limitations is useful, many op-amp circuits work perfectly well without any consideration of the following limitations. There will always be the odd circuit that will need special consideration. This chapter covers the awkward few.

When you buy an op-amp, you will have selected it from a list of many others. How do you know which one to choose? Investigations will lead you to long lists of specifications for each of the different chips and you will need to have some grasp of the meaning of these specifications. Let us look at some fundamental *real* op-amp specifications, their names and what they mean.

5.1.1 Input bias current

For an ideal op-amp the inputs draw no current. Now for a real op-amp, the inputs *do* have a small amount of current flowing into (or out of) them. The reason for this small but finite current flow is that the inputs to the op-amp drive some transistor devices. We have treated the op-amp as a black box so far. Let us not pretend that you will *never* need (or want) to know what is inside it (usually you do not): now is the time for a small peek into the box.

Look at Figure 5 1. The input to the op-amp chip looks a bit like this. It is not important for you to understand transistor action just now, but only for you to see that I_{bias} is a small current that flows all the time that the op-amp is powered up, and its purpose is to set up the conditions (known as the 'biasing') required by the first transistor in the circuit. The op-amp has one of these at *each* of its two inputs and manufacturers do all they can to make the input transistors identical (called 'matching') Unfortunately, the two transistors always differ slightly, which means that the I_{bias} into one input is always slightly different from the I_{bias} into the other. To a certain degree of accuracy they could be considered to be equal. This leads us to a definition for the real op-amp, called the 'input bias current'. Because of the

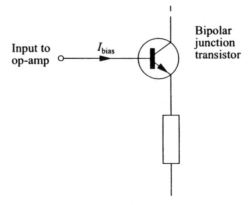

Input to op-amp I_{bias} Bipolar junction transistor

Figure 5.1

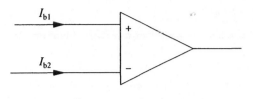

Figure 5.2

slight difference in the two input currents, we quote an average of the two currents and call it I_{bias}. In Figure 5.2 you can see that each input has its own current flow· I_{b1} or I_{b2}. We take the average of these to give us our bias current value. Therefore

$$I_{bias} = (I_{b1} + I_{b2})/2$$

There is no significance in the numbering of these currents; I_{b1} could just as easily be shown flowing into (or out of) the inverting input. The important part of this information is that I_{bias} is small (a few nanoamps) and we can choose to neglect it or not It depends on the circuit, as we will see shortly.

One final point about I_{bias} is that it flows *into* some op-amps (ones containing npn and n-channel devices) and *out of* others (the pnp and p-channel devices). Our way of dealing with any I_{bias} problems is the same for both types. For a typical 741 value

$$I_{bias} \approx 80 \text{ nA}$$

This value is approximate and tells us to expect 80 nA, or less, for any 741 in use. You can see that this is a small current, and negligible for most considerations. (Note how we treated it in our ideal op-amp analysis) It is important to note that the I_{bias} is a steady d.c current. The next specification is closely related to I_{bias}.

5.1.2 Input offset current

We have just mentioned that the two input currents to the op-amp differ owing to the mismatch between the input transistors The input offset current ($I_{in(off)}$) is simply the difference in these currents Therefore, we define it as

$$I_{in(off)} = |I_{b2} - I_{b1}|$$

Note here the use of the modulus sign. This is used because we have no idea which current is the larger, but the $I_{in(off)}$ is always a positive value For a typical 741 value

$$I_{in(off)} \approx 20 \text{ nA}$$

This value is approximate and tells us to expect 20 nA, or less, for any 741 in use
Now we look at voltages at the input

5.1.3 Input offset voltage

First, let us consider what we would expect of an idea! op-amp if we connected its inputs together (see Figure 5.3) We have a situation where the voltage across the input of the op-amp is a true zero. This is *not* the same as our rule 1 approximation: that the input voltage is negligibly small. This time we really *do* have a zero input voltage. What will the output be? The output can only be zero. With the best will in the world we cannot amplify *nothing* to produce *something*, however much voltage gain we have! Unfortunately, in the world of real op-amps the answer is not so simple. The mismatched input transistors not only cause the input offset current. they also conspire to produce small unbalanced input voltages within the input circuitry of the device. Because of the very high voltage gain of the op-amp, any voltage mismatch of this sort is amplified and produces an unwanted output voltage.

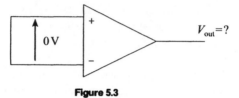

Figure 5.3

The input offset voltage $V_{in(off)}$ is a slightly odd parameter as it represents the size of an external voltage source that would need to be added to one of the input circuits to make the output zero when the inputs are shorted together! This idea is represented in Figure 5.4, where the voltage source is drawn in one of the inputs. For a typical 741 value

$$V_{in(off)} \approx 2 \text{ mV}$$

Although values are quoted for this parameter, we have little real knowledge of it, since we are never sure of its polarity or which one of the inputs requires it. All we know is that the quoted value is approximate and tells us to expect 2 mV or less for any 741 in use. Do not let this parameter worry you unduly. It seems at first sight to be rather odd and quite large. In general circuit operation, however, its

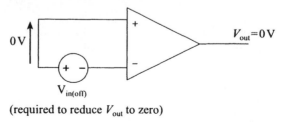

(required to reduce V_{out} to zero)

Figure 5.4

effect is quite minimal Manufacturers often provide easy ways of reducing its effect anyway. (With the 741, for instance, we connect a $10\,k\Omega$ potentiometer between pins 1 and 5 with the wiper connected to the negative supply rail.)

It is important to note that $V_{in(off)}$ is a d c. voltage

5.1.4 Open loop voltage gain

You have met the open loop voltage gain (A_{vol}) before What else do we have to say about it? The approximate A_{vol} for any specific op-amp type is quoted as the voltage gain at d.c. A typical 741 value is 200 000 However, when the input is a c. (we start to oscillate the input at some periodic frequency), the amplifier does not amplify at the same constant value It is okay for low frequencies (as for d.c. input), but as we increase the input frequency the A_{vol} of the op-amp starts to drop. In fact, for a 741 the gain drops to only 1000 (!) if the input frequency is 1000 Hz. If we keep on increasing the frequency, we actually reach a point where the A_{vol} is only 1. This frequency is often called the 'unity gain frequency', f_T. (You might think that this sounds like a large deviation from an ideal op-amp and you would be correct: it certainly is!) Figure 5 5 is the typical graph of the change in A_{vol} with frequency for a 741. Because of the large range of values involved, it is common to take logs of both scales

You can see that A_{vol} reduces linearly with frequency above approximately 10 Hz; this is called 'frequency roll-off' For the 741 $f_T = 1$ MHz. What will seem strange is that this frequency response is *intentionally* designed into the op-amp; it is known as 'frequency compensation'. You might reasonably ask why its performance is limited intentionally The simple answer is that the manufacturers

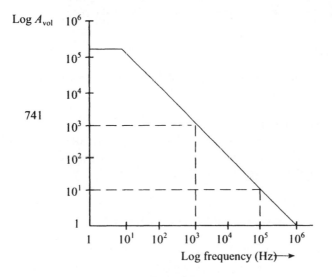

Figure 5.5

build frequency compensation into the circuitry to increase stability. Without it, it is quite possible that standard op-amp circuits may start oscillating and become unstable. It is basically to do with phase changes through the chip at higher frequencies, causing negative feedback to turn into positive feedback! It means that a frequency-compensated op-amp will work well in all standard configurations. (If an uncompensated op-amp is chosen, it is the circuit designer's task to ensure its stability by adding extra components into the circuit.)

This gain/frequency response curve is quite helpful in determining the useful frequency range of a circuit (see Section 5.1.5 on bandwidth). You will notice in Figure 5.5 that if we draw a horizontal line at a gain of 10, the intercept with the response curve is at a frequency of 100 kHz. This tells us that we cannot achieve any greater gain than this with *any* standard op-amp circuit configuration at 100 kHz. It would be no use, therefore, in using this op-amp as (say) a non-inverting amplifier with external resistors chosen to give A_v at 20 if we cannot achieve 20 in the open loop circuit. Notice also that here A_v means the voltage gain of the complete circuit, not just the op-amp. A_v is sometimes called A_{vcl} or 'closed loop voltage gain' to indicate the difference in gain with a negative (or closed) feedback loop.

What we learn from this is that we must be aware of the upper limit in overall voltage gain of any of our circuits when we start using higher frequencies. The trick is to keep A_v well below the A_{vol} figure at the highest frequency you intend to use with the circuit

EXAMPLE 5.1
If we design a 741 inverting amplifier to operate up to 1 kHz, we must keep its A_v below -1000 ($A_{vol(1\ kHz)} = 1000$). A good rule would be to attempt 10% of the A_{vol} value. In this case, make $A_v = -100$ and no more!

5.1.5 Bandwidth

The upper frequency limitation or useful frequency range mentioned in Section 5.1 4 is indicated by the parameter known as the 'bandwidth' (BW) of the circuit. Figure 5.5 shows that the highest A_{vol} value is constant from d.c. to approximately 10 Hz. At this gain, therefore, the useful range of frequencies is 0–10 Hz. Therefore, the bandwidth is 10 Hz when the gain is approximately 200 000 The bandwidth for an op-amp is found by reading off the frequency value at the point on the graph where the required gain crosses the response curve

EXAMPLE 5.2
What is the bandwidth of the 741 at a gain of 1?

SOLUTION
From Figure 5 5 you can see that the bandwidth at this gain is 1 MHz.

It soon becomes obvious that there is a relationship between the gain and the BW. We look at this next

5.1.6 Gain bandwidth product

The relationship between BW and gain is that their product will always be constant for a specific op-amp The constant is called the 'gain bandwidth product' (GBWP), i.e.

$$BW \times gain = GBWP$$

EXAMPLE 5.3
What is the GBWP of the 741?

SOLUTION
We can take any pair of corresponding A_{vol} and BW values. We will choose $A_{vol} = 1000$ and BW = 1 kHz Therefore

$$GBWP = 1000 \times 1000 \text{ Hz} = 1 \text{ MHz}$$

An important point here is that the A_{vol}/frequency response curve can also be used to measure the BW and GBWP of any closed loop circuit.

EXAMPLE 5.4
What is the bandwidth of the 741 op-amp voltage follower circuit?

SOLUTION
We recall that the closed loop gain A_v for the follower is 1 and we know that the 741 GBWP is 1 MHz Therefore, since

$$GBWP = A_v \times BW$$

$$BW = GBWP/A_v$$

$$= 1 \text{ Mhz}/1$$

$$= 1 \text{ MHz}$$

Once we know the GBWP of a certain op-amp we can find the bandwidth at any frequency without having to use the response curve

EXAMPLE 5.5
What is the bandwidth of a 741 inverter circuit with $A_v = -10$ at 200 Hz?

SOLUTION

We know that for a 741 the GBWP is 1 MHz. Therefore

$$BW = GBWP/A_v$$

$$= 10^6 \text{ Hz}/10$$

$$= 10^5 \text{ Hz}$$

$$= 100\ 000 \text{ Hz}$$

5.1.7 Slew rate

When we drive an op-amp inverting or non-inverting amplifier with an a c. signal we expect the shape of the output to be a faithful reproduction of the shape of the input signal. A problem associated with real op-amps is the inability of an op-amp to be able to change its output voltage quite as quickly as we would sometimes like. This rate of change of output voltage has a maximum value for a specific op-amp and is called the 'slew rate' (SR). In calculus it would be represented as the dV/dt value. It is quoted in either volts per second (V s^{-1}) or volts per microsecond (V μs^{-1}).

Usually the reason why the output is limited in this way is because of the internal frequency compensation capacitor. The most obvious sign in practical circuits that the output has become slew rate limited is when a sine wave output starts to take on a triangle wave look with a reduced amplitude. For a typical 741 value

$$SR \approx 0.5 \text{ V}\mu\text{s}^{-1}$$

EXAMPLE 5.6

We require a 741 op-amp output to give a 10 V$_{pk-pk}$ triangle waveform at 50 kHz. Is the circuit capable of producing this?

SOLUTION

The period of the wave T is 1/50 kHz = 20 μs A sketch of the waveform (Figure 5 6) shows us that the output must change 10 V in 10 μs, i.e. 1 V every 1 μs. Unfortunately, the 741 can only give a 0.5 V change every 1 μs It is therefore unable to give us the required waveform The result is an output waveform that is changing as fast as it can but, as you can see in Figure 5.6, its slope is reduced and therefore cannot reach the required voltage before it is sent back in the other direction

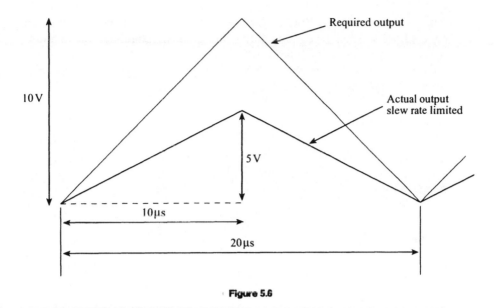

Required output

Actual output
slew rate limited

10 V

5 V

10 µs

20 µs

Figure 5.6

When the output is distorted in any way by this effect, we call the result a 'slew rate limited output'.

5.1.8 Output saturation levels

When a real op-amp is used in any circuit there is a limitation on the size of output voltage available from it We must remember that the op-amp is a complex integrated circuit and requires a power supply to drive it, just like any other electronic black box we use. The size of the power supply voltage imposes a limit on the maximum available output voltage from the op-amp. For many different types of op-amp, the maximum output voltage is just within the power supply voltage and we must expect to lose up to a volt inside the chip. Most op-amps have two power supply pins: one positive and one negative. The positive pin receives a positive voltage and the negative pin can be connected to a negative voltage or earth (0 V). A bipolar supply means using both positive and negative supply voltages, and allows the op-amp output to swing positively and negatively. Having the negative pin held at earth is called a 'single-ended supply' and prevents the output from going negative.

EXAMPLE 5.7
What will be the maximum output voltages available with a 741 circuit powered from a +/−15 V supply?

SOLUTION
You will find that the output voltage maximum values are approximately −14 V to +14 V

When a circuit that drives the op-amp output to one of its output saturation levels is used, we say that the op-amp is 'saturating' or 'in saturation'. Saturation is a condition we try to avoid in all linear circuits, since an output pushed and held at its saturation level usually means that it is not able to give us the required output and we lose the information.

EXAMPLE 5.8

A 741 inverter circuit, powered at +/−15 V, has a gain of −20 and a 50 Hz sine wave $5V_{pk}$ input Will the output reach saturation?

SOLUTION

$$V_{out(pk)} = A_v \times V_{in(pk)}$$

$$= -20 \times (+/-5)$$

$$= +/- 100 \text{ V}$$

This is much larger than +/−14 V; therefore, the output will saturate Its effect will be to remove (or 'clip') the positive and negative peaks of the inverted output sine wave. It will look very much like a 50 Hz, $28V_{pk-pk}$ square wave with sloping sides rather than vertical sides. Try drawing it for yourself to get the idea

So remember, with practical circuits it is pointless designing any circuit that is trying to deal with a voltage that could theoretically exceed any of the power supply voltages, unless that is the designer's intention (see Chapter 6 on switching circuits) This limitation on the size of saturated output voltage level is dependent on the load current drawn from the op-amp output, i.e. if the load current is high, the saturated voltage to the load is reduced slightly.

If you remember that the op-amp output can be represented as a Thevenin circuit with an internal resistance, you can hopefully understand why a larger load current will reduce the available output voltage from the op-amp. (The op-amp output is a good voltage source, but not perfect!) We are generally talking about load currents of about 10–20 mA.

We have now covered a few of the real op-amp limitations a designer may need to know about, so before we move on to look at some circuits, let us try a little self-assessment

Self-assessment questions

SAQ5.1 What is the bandwidth of a 741 op-amp inverter circuit with A_v = −20?

SAQ5.2 What is the open loop gain of a 741 at 5 kHz, 50 kHz and 500 kHz?

SAQ5.3 What would be a sensible maximum value of A_v for a 741 inverter at 10 kHz?

SAQ5.4 Why would there be a problem if an inverter were built from a 741 with a feedback resistance of 1000 kΩ, an input resistance of 1 kΩ and an input signal frequency of 5 kHz?

SAQ5.5 What is the name and symbol for the frequency at which A_{vol} is 1?

SAQ5.6 The input currents to a particular op-amp are measured to be 10 nA into one input and 20 nA into the other. What are I_{bias} and $I_{in(off)}$ for the op-amp?

SAQ5.7 If an op-amp is quoted as having $I_{bias} = 100$ nA and $I_{in(off)} = 20$ nA, what are the input current values?

SAQ5.8 The I_{bias} for a certain op-amp is 30 nA. Is the larger of the two currents flowing in the inverting or in the non-inverting terminal?

SAQ5.9 What part of a sine wave has the maximum rate of change of voltage?

SAQ5.10 If a 20 V_{pk-pk} triangle wave of 30 kHz is used as input to an op-amp voltage follower circuit whose slew rate is 1 V μs^{-1}, would the output be slew rate limited?

SAQ5.11 If a 10 V_{pk-pk} sine wave of 20 kHz is used as input to a 741 op-amp voltage follower circuit whose slew rate is 0.5 V μs^{-1}, would the output be slew rate limited? (Hint· this involves the need to differentiate $V_{pk} \sin(\omega t)$.)

SAQ5.12 A 741 op-amp non-inverter is powered by a bipolar +/−10 V supply. It is required to amplify a 1 V_{rms} sine wave at 1 kHz. What is the approximate maximum gain possible before saturation and clipping of the output occur?

5.2 Circuit design modifications

We now look at *real* inverting and non-inverting circuits in order to demonstrate where and when certain circuits need to be amended because of non-ideal op-amp behaviour

5.2.1 The inverting amplifier

The ideal circuit is perfectly okay to use for most applications, but for some cases I_{bias} may be a problem. Look at the circuit in Figure 3.6. The I_{bias} for the real op-amp *must* be able to flow into (or, for some circuits, out of) the op-amp input, otherwise the op-amp will not work. Where does I_{bias} come from? Well, it must be supplied by the current paths through R_1 and R_2. So now let us assume that V_{in} and V_{out} come from ideal voltage sources This can then be simplified to find the Thevenin resistance (Remember, short the voltage sources and see which resistance network is left.) We have the situation shown in Figure 5 7. This leaves R_1 between the inverting input and earth, which is where R_2 also is, i.e R_1 is in parallel with R_2 between the earth and the inverting input

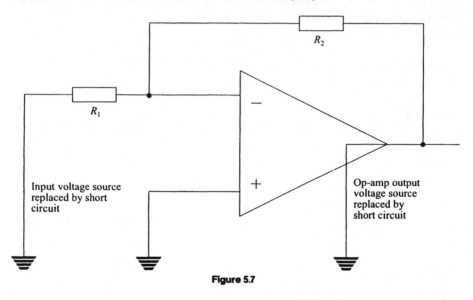

Figure 5.7

Now we can see where I_{bias} comes from it is drawn up from earth through the parallel connection of R_1 and R_2. I_{bias} flowing through R_1 and R_2 in parallel causes an unwanted voltage drop to appear at the inverting input This voltage drop is called an 'offset voltage'. The resulting offset voltage at the inverting input is given the term $V_{-(os)}$ and the size of this offset is given by Ohm's law

$$V_{-(os)} = I_{bias}[R_1 \times R_2/(R_1 + R_2)]$$

This voltage is amplified by the A_{vcl} (closed loop gain) action of the circuit and causes a larger offset output voltage to be created. The unwanted output offset voltage is given the term $V_{o(os)}$. Therefore, we can say

$$V_{o(os)} = V_{-(os)} \times A_{vcl}$$

The problem created is significant only if I_{bias} or A_{vcl} is large. The op-amp's internal input offset voltage $V_{in(off)}$ is also amplified by the circuit action, but we are never sure whether it is positive or negative, and therefore its contribution is unknown If we are in doubt we simply add its value to $V_{-(os)}$, which gives a total input offset voltage called $V_{i(os)}$

$$V_{i(os)} = V_{in(off)} + V_{-(os)}$$

Therefore

$$V_{o(os)} = V_{i(os)} \times A_{vcl}$$

Its effect is to give a non-zero output when V_{in} to the circuit is zero.

EXAMPLE 5.9
We want to use an op-amp with $I_{bias} = 50\,nA$ and $V_{in(off)} = 1\,mV$ in an inverter with $A_v = -10$. We need an input resistance of at least $100\,k\Omega$. What will be the size of the output offset voltage $V_{o(os)}$?

SOLUTION
We must choose the R_1 and R_2 values. We will make $R_1 = 100\,k\Omega$ and $R_2 = 1\,M\Omega$ in order to give the required A_v and to satisfy the R_{in} requirement. Now

$$R_1 \parallel R_2 = 100\ k\Omega \times 1\ M\Omega/(100\ k\Omega + 1\ M\Omega)$$
$$= 90.9\ k\Omega$$

Therefore, the voltage drop is

$$V_{-(os)} = 50\ nA \times 90.9\ k\Omega$$
$$= 4.55\ mV$$

Including $V_{in(off)}$ gives

$$V_{i(os)} = 1\ mV + 4.55\ mV$$
$$= 5.55\ mV$$

This is increased by the gain to give

$$V_{o(os)} = 5.55\ mV \times (-10)$$
$$= 55.5\ mV$$

Not too much of a problem!

However, if the A_v required was -500 and I_{bias} was $200\,nA$, $V_{o(os)}$ would increase to above $9\,V$ and become a problem. Recall that we have not yet considered adding a signal to the circuit!

Self-assessment question

SAQ5.13 Work out the value of $V_{o(os)}$ in the second case of Example 5.9

Fortunately, there is a simple technique for reducing this unwanted output voltage. All that is required is another resistance in the non-inverting input circuit equal in value to $(R_1 \times R_2)/(R_1 + R_2)$ This quick-fix method equalizes the resistance paths for the bias current in both input circuits. Since the bias current and the resistance is equal in both, they both therefore have the same offset voltages created at their inputs, which merely cancel each other out. In Figure 5 8 the R_3 resistance is chosen to be approximately $R_1 \parallel R_2$.

This method might not completely cancel the output offset, but it certainly reduces it to a negligible size.

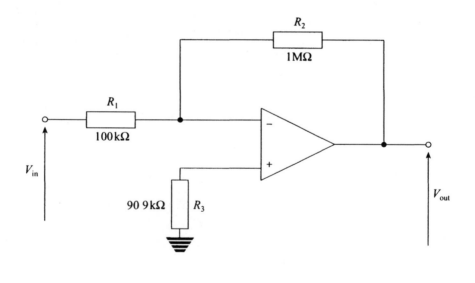

Earth

Figure 5.8

Self-assessment question

SAQ5.14 An inverter with an I_{bias} of 400 nA is designed to give a gain of −200 Choose appropriate resistance values to provide this and then (using the values you have chosen) find the $V_{o(os)}$ created. Finally, add a circuit amendment to counteract the offset created. Draw the circuits.

5.2.2 The non-inverter

We now investigate similar effects on the non-inverting circuit The approach is similar to that for the inverting amplifier. Look at Figure 3.8 and consider the following action.

First we find the Thevenin resistance between the inverting input and earth. Remember to short out any voltage sources. Do you see that we again have R_1 and R_2 in parallel. These resistances create an input offset voltage at the inverting input $(V_{-(os)})$ when the bias current flows and the output offset created is an amplified version of it. The only difference between this and the inverter circuit is that the gain equation is slightly different The solution is to insert an extra resistance R_3 in the circuit to the non-inverting op-amp input in order to equalize the resistances seen by both the I_{bias} currents (see Figure 5 9). The resistance added is equal to the parallel combination of R_1 and R_2 Just as in the case of the inverter circuit, this problem is of significance only if A_{vcl} or I_{bias} is particularly large

Something particular to watch out for is when a non-inverter circuit is driven by a poor voltage source V_s with a high internal resistance R_s. I_{bias} will cause a d c. offset $V_{+(os)}$ to appear at the non-inverting input

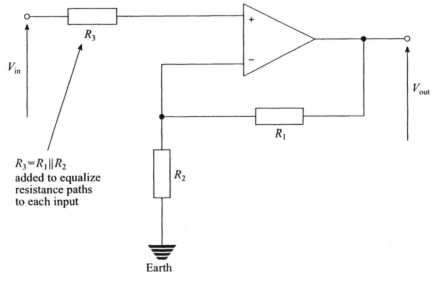

$R_3 = R_1 \| R_2$
added to equalize
resistance paths
to each input

Figure 5.9

$$V_{+(os)} = I_{bias} \times R_s$$

We cannot easily deal with this because R_s is internal to the source and therefore untouchable. All we can do is to calculate the total input offset voltage $V_{i(os)}$ to the op-amp and hope that it is not too large to cause problems

$$V_{i(os)} = V_{in(off)} + |(V_{+(os)} - V_{-(os)})|$$

The modulus of the difference in the offsets at each input is used The output offset is then found from

$$V_{o(os)} = V_{i(os)} \times A_{vcl}$$

as before.

The best solution, of course, is to place a voltage follower between the source circuit and the non-inverter amplifier. The voltage to the non-inverter then comes from a good voltage source!

Self-assessment question

SAQ5.15 If a non-inverter circuit (for a.c. or d c.) has a gain of 20, a feedback resistance of 57 kΩ and is driven by a signal source of 25 kΩ output impedance·
1. Draw the circuit.
2 What size is the resistance from inverting input to earth?
3 If both op-amp inputs require the same bias current, which input will have the largest input offset? (Remember that in this case the non-inverting input must draw its current from the signal voltage source.)
4. Assuming that we calculated $V_{-(os)} = 1.5$ mV and $V_{+(os)} = 12 5$ mV for a circuit of this kind and, given $V_{in(off)} = 2$ mV for this op-amp, find $V_{i(os)}$ and $V_{o(os)}$

5.2.3 The a.c. non-inverter

To enable a circuit to work with a.c signals only, we usually merely insert a capacitor at the input and output of the circuit to block the d c voltage or current levels from one another so that any d c. output voltage or current from a source circuit will not affect the d.c. levels of a load circuit. So the a.c. non-inverter circuit will look like the circuit in Figure 5.10.

Unfortunately, this circuit will never work. Why? Well, remember that the small (but necessary) I_{bias} current *must* flow Where does I_{bias} for the non-inverting input come from? In previous circuits the current was drawn from earth and into the input or up through a resistance and into the input. However, in this circuit no such thing can happen because the capacitor blocks any d.c. current flow and therefore the op-

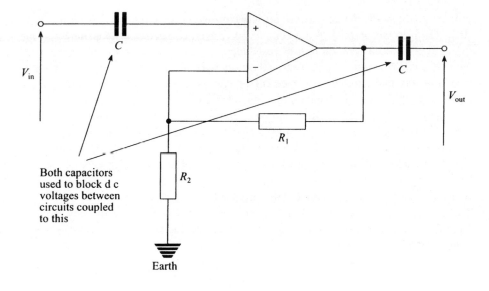

Both capacitors
used to block d c
voltages between
circuits coupled
to this

Earth

Figure 5.10

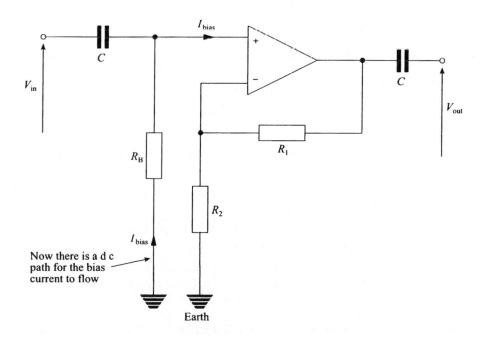

Now there is a d c
path for the bias
current to flow

Earth

Figure 5.11

amp does not work To get the circuit working we *must* make provision for I_{bias} to flow. How do we do this? We add an extra resistance between the ground and the input. It is called a 'ground return' resistance. This is shown as R_B in Figure 5.11. Now I_{bias} can flow through it. That is the *only* purpose of the resistance Of course, it also means that another unwanted offset voltage is created across R_B! Make R_B equal to $R_1 \parallel R_2$ and we balance this out. Unfortunately, if R_B is made too small, the input resistance R_{in} to the complete circuit becomes dominated by its small value. This would be a shame since one of the redeeming features of the non-inverting circuit is its high R_{in} value! In fact, it's another trade-off between good voltage transfer and small input offset. (100 kΩ is often used as a general rule.)

A circuit trick to stop $V_{I(os)}$ from being amplified

These offset problems are all d c. quantities and since we want only an a.c. amplifier we can add an amendment to this circuit which (rather cleverly!) gives it unity gain at d c., but full gain at a.c. Look at Figure 5.12 and notice the capacitor C_3 What is its purpose?

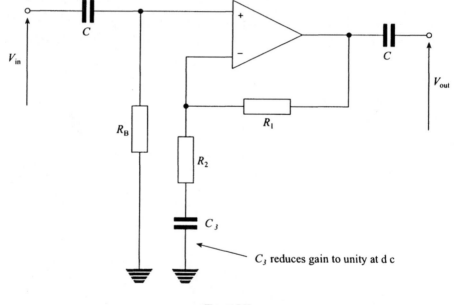

Figure 5.12

To understand the purpose of C_3 here, we need to recall how the reactance X_c of a capacitor varies with frequency Since $X_c = 1/2\pi fC$, we can see that X_c is infinite at d.c. when $f = 0$ Hz and has a very low value at a.c when f is very high.

If we therefore choose a value of C which gives an almost negligible X_c value at all the frequencies we need, we will find that the bottom end of R_2 is effectively connected to earth at a c. This is often called an 'a.c. earth' and is sometimes used in other circuits. Therefore, at a c.

$$A_{vcl} = 1 + (R_1/R_2)$$

as normal. This will amplify the incoming a.c. waveform.

For the d.c. response, X_c will be an infinite value and we can see that C_3 will appear to act as an open circuit. The total resistance (or impedance) between the inverting input and earth is now infinity. This alters the gain equation for the circuit to

$$A_{vcl} = 1 + (R_1/\infty)$$
$$= 1 + 0$$
$$= 1$$

i e gain is unity at d.c. In other words, the circuit is turned into a voltage follower at d c. The beauty of the arrangement is that all the nasty input offset voltages we were considering are amplified by only 1 They are no larger at the output than they are at the input. Wonderful! This is an easy quick-fix to an otherwise tricky problem

5.2.4 The integrator

Look for a moment at the ideal op-amp integrator circuit shown in Figure 4.5 which we analysed earlier using the ideal op-amp rules. With reference to what we now know about *real* op-amps, we can now investigate another effect here. Imagine for a moment that the V_{in} is held at earth. Remember I_{bias}? It flows through R to bias the op-amp circuit. You know now that this will create an offset voltage ($I_{bias} \times R$) at the inverting input ($V_{-(os)}$). But the circuit is an integrator! Therefore, the small offset voltage is integrated over a time interval which starts whenever the circuit becomes powered up. The resulting output is a rising (or falling) ramp voltage which increases linearly with time. Does it rise to infinity? No. It saturates at the power rail level. This means that any op-amp integrator will saturate before you have a chance to give it an input! One explanation for this is that the integrator can be viewed as a simple inverter circuit with a reactance feedback instead of a resistance feedback. Instead of $A_v = -R_2/R_1$, we can write

$$A_v = -X_c/R_1$$

Remember that reactance is frequency-sensitive and is large at low frequencies. At d c , X_c is infinitely large. This means that we have infinite gain for our d.c. offset Not good! We prevent this from happening by limiting its gain to ten or so at d c. by adding a feedback resistance R_f ten times larger than R across the capacitor, i e. $R_f = 10R$ (see Figure 5.13). The effect of this is merely to turn the circuit into a -10 inverter at d c. and low frequencies.

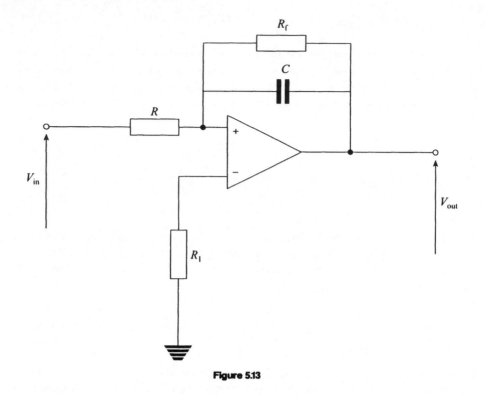

Figure 5.13

The circuit now works happily at a.c. and we normally choose C to have a reactance of less than R_f at all the frequencies we want to use the circuit with. This means that R_f becomes dominant only at d.c. and at frequencies lower than those being used.

In the usual way a resistor R_1 is inserted in the non-inverting input circuit to try to balance out the input offset voltages, its value being approximated to $R \parallel R_f$.

Checklist

Do you know:
- the meaning of op-amp bias and offset currents?
- the meaning of op-amp input and output offset voltages?
- what a slew rate limited output is?
- how bandwidth is affected by gain?
- what a saturated output means?
- how to reduce the input offset of the standard inverting and non-inverting circuits?
- why a ground return resistance is sometimes required?
- why an integrator circuit does not work very well at d.c ?

Tutorial questions

1 What is the bandwidth of a 741 op-amp inverter circuit whose A_v is -50?

2 What is the A_{vol} of a 741 at 1 kHz, 10 kHz and 100 kHz?

3 What would be a sensible maximum A_v for a 741 inverter at 20 kHz?

4 If an inverter were built from a 741 with a feedback resistance of 150 kΩ and an input resistance of 1 kΩ, what would be a sensible maximum input signal frequency?

5 What does f_T stand for?

6 The input currents to a particular op-amp are measured to be 15 μA into one input and 45 μA into the other What are I_{bias} and $I_{in(off)}$ for the op-amp?

7 Is the GBWP constant for a 741 op-amp?

8 If an op-amp is quoted as having $I_{bias} = 0.1$ nA and $I_{in(off)} = 30$ pA, what are the input current values?

9 Is it true that the non-inverting input always draws the largest current?

10 What part of a square wave has the maximum rate of change of voltage?

11 If a voltage follower circuit, driven by a 10 V_{pk-pk} triangle wave, has SR = 10 V μs^{-1}, what is the highest possible frequency that could be used before slew rate limiting starts?

12 If a triangle wave with $V_p = 5$ V and frequency 30 kHz is used as input to an op-amp voltage follower circuit with SR = 0.5 V μs^{-1}, is the output slew rate limited?

13 A 6 V_{pk-pk} sine wave of 10 kHz is used as input to an op-amp voltage follower circuit with SR = 0 1 V μs^{-1}. Is the output slew rate limited?

14 A 741 op-amp inverter is powered by a bipolar +/−12 V supply. It is required to amplify a $5V_{rms}$ sine wave at 1 kHz. What is the approximate maximum gain possible before saturation and clipping of the output occur?

15 We want to use an op-amp with $I_{bias} = 1$ μA and $V_{in(off)} = 2.5$ mV in an inverter with $A_v = -250$. We need an input resistance of at least 25 kΩ. What will be the size of the output offset voltage $V_{o(os)}$?

16 An inverter with $I_{bias} = 800$ nA is designed to give a gain of -100. Choose appropriate resistance values to provide this, then (using the values you have chosen) find the $V_{o(os)}$ created Finally, add a circuit amendment to counteract the offset created. Draw the circuits.

17 A non-inverter circuit (for a c or d c) has a gain of 50, a feedback resistance of 500 kΩ and is driven by a signal source of 100 kΩ output impedance:
   - Draw the circuit.
   - What size is the resistance from inverting input to earth?
   - If both op-amp inputs require the same bias current, which input will have the largest input offset?

18 A non-inverting circuit of $A_{vcl} = 200$ produces offsets of $V_{-(os)} = 105$ mV

and $V_{+(os)} = 25\,\text{mV}$. Find $V_{i(os)}$ and $V_{o(os)}$. You may assume $V_{in(off)} = 2\,\text{mV}$ for the op-amp

19   What is the purpose of a ground return resistor?

20   What is meant by an 'a c. earth'? Give an example of its use.

Further reading

Bogart T F (1986) *Electronic devices and circuits*, sections 13 3–5. Merrill, Columbus, Ohio

Floyd T L (1988) *Electronic devices*, 2nd edn, pp 459–70 Merrill, Columbus, Ohio

Horowitz P & Hill, W (1987) *The art of electronics*, sections 3 11–12 Cambridge University Press, Cambridge, England

CHAPTER 6

Op-amp switching circuits

Aims and objectives

The aims of this chapter are to explain briefly what a switching circuit is and to take a look at the most popular of the op-amp switching circuits in use. We will see how circuits can easily be designed to turn things on and off, as in timers, alarm systems, thermostats, light-operated and sound-operated equipment and so on. We will also investigate briefly the use of positive feedback and the design of a simple square wave oscillator circuit

When you have studied this chapter you will have a better understanding of:

- *simple voltage comparators*
- *resistance monitoring circuits*
- *delay timers*
- *window comparators*
- *positive feedback*
- *Schmitt triggers*
- *astable oscillators*

All the circuits we have looked at so far are called 'analog circuits'. This means that they have all been designed to give outputs which can take an infinite number of different values between two voltage levels. These outputs are sometimes called 'continuous functions' and their output waveforms could be drawn without having to remove pen from paper.

Switching circuits, however, involve output levels which change almost instantaneously from one voltage level to another and do not give continuous output functions. As far as op-amp circuits go, this means that the op-amp output switches from one output saturation level to the other. This type of circuit forms the link from analog circuits to digital circuits. Digital circuits use only two voltage levels· on and off, or high and low All switching circuits are also non-linear circuits, since an output which is saturated cannot respond to any further small changes in input.

If an op-amp output is forced into saturation, we unfortunately can no longer apply both of our two easy op-amp rules. In fact, it is rule 1 that is no longer applicable· the rule that tells us that the input voltage difference across the op-amp is negligibly small. In switching circuits the input voltage difference can become rather large, forcing the output heavily into one of its saturation levels Fortunately, rule 2 still applies, which means that we can still analyse switching circuits without too much hard work.

6.1 The basic comparator

First, a comparator must compare things, so what are we comparing? Well, the circuit is used to compare two input voltage levels. It then indicates which of the voltages is the larger, by providing one of two output voltage levels. Quite simple really! If the op-amp is powered up and used in its basic open-loop form (no feedback circuitry), it becomes a comparator automatically. The action of the comparator is best explained by an example, since its action is partly dependent on the type of op-amp used.

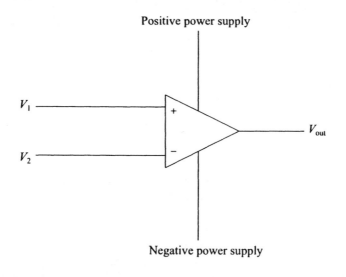

Figure 6.1

EXAMPLE 6.1
Referring to Figure 6 1, consider we have a 741 and we use a +/−15 V power supply. Now, what happens if
1   $V_1 = 4$ V and $V_2 = 3$ V?
2. $V_1 = 4$ V and $V_2 = 5$ V?

SOLUTION

We remember that $A_{vol(741)} = 200\,000$. This means that the tiniest difference in voltage at the input will be amplified by $200\,000$ We also remember that the output saturation levels are approximately $+/-14\,\text{V}$.

1   In this case there is a voltage difference V_{diff} across the inputs of 1 V. The non-inverting input is 1 V more positive than the inverting input. Therefore, the output will go positive; but to what value? Using the gain equation

$$A_{vol} = V_{out}/V_{diff}$$

Therefore

$$V_{out} = A_{vol} \times V_{diff}$$

$$= 200\,000 \times (+1 \text{ V})$$

$$= +200\,000 \text{ V}$$

From what we know about real op-amps, we can see that V_{out} cannot exceed its positive saturation level. Therefore, although the output is trying to reach $+200\,000\,\text{V}$, it bangs its head on the positive supply rail voltage. We get $V_{out} = +14\,\text{V}$ or thereabouts

2.   Here we also have a 1 V difference voltage, but now the inverting input is more positive than the non-inverting input. The output therefore goes negative

$$V_{out} = 200\,000 \times (-1 \text{ V})$$

$$= -200\,000 \text{ V}$$

Of course, it is limited by the power rail voltage, giving $V_{out} = -14\,\text{V}$.

EXAMPLE 6.2

Refer again to Figure 6 1. What is the smallest V_{diff} value that will result in saturation at $V_{out} = +14\,\text{V}$?

SOLUTION

We know that

$$A_{vol} = V_{out}/V_{diff}$$

Therefore

$$V_{diff} = V_{out}/A_{vol}$$
$$= +14/200\,000$$
$$= +70\mu V$$

This is a very small V_{diff}; we would similarly find that $-70\,\mu V$ results in a -14 V output

What we can learn from these examples is that the output can only have one of two possible values Each output value represents a particular input condition which is a comparison of the sizes of the input voltages at each input. The different input conditions are compared in Table 6 1 Hey presto! We have a circuit which compares two voltages and gives us an answer (of sorts).

Table 6.1

Input condition	Output value	Interpretation of the output voltage
If $V_1 > V_2$	High ($+14$ V)	V_1 is higher in voltage than V_2
If $V_1 < V_2$	Low (-14 V)	V_1 is lower in voltage than V_2

Applications of a circuit like this are numerous, especially when we consider that the comparison can be made more flexible if we fix one of the inputs to a known reference voltage. Figure 6.2 shows the comparator as a heat sensor.

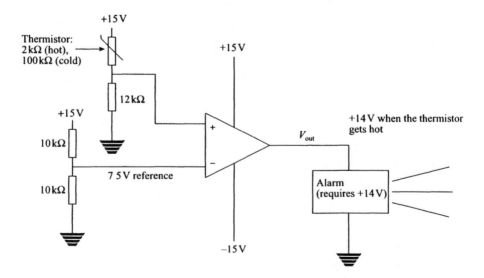

Figure 6.2

6.2 The comparator as a temperature sensor

The circuit in Figure 6 2 uses a voltage divider to create 7.5 V as the reference and gives a high output whenever the thermistor (which is a heat-sensitive resistance) is placed in a hot environment This could be used to detect the temperature of a hot-water pipe or a room The output is used to drive an indicator light, an alarm or other similar device.

Self-assessment question

SAQ6.1 What will be the voltage at the non-inverting input when the thermistor is:
1.   cold,
2   hot?

You might notice that if one of the reference resistances was made variable we could introduce more control of the temperature which would set off the alarm.

Self-assessment question

SAQ6.2 What would the circuit do if the positions of the thermistor and the 12 kΩ resistor were interchanged?

The number of variations on this type of circuit is rather large. We could use light-sensitive, pressure-sensitive or humidity-sensitive resistors, or anything whose resistance is sensitive to some environmental change.

6.3 The comparator as a delay timer

The circuit for a delay timer is shown in Figure 6.3. It compares the reference voltage V_{ref} (7.5 V) with the voltage across a charging capacitor V_c in a series RC circuit. When the reset switch SW1 is opened, the circuit starts timing and V_c rises exponentially towards the target voltage V_{target} of +15 V. When V_c reaches and surpasses V_{ref}, the output voltage switches from -14 V to +14 V. This means that the output switches after a fixed time delay. The capacitor continues to charge towards its target of 15 V. The circuit will stay in this condition until SW1 is closed (discharging the capacitor) and reopened, allowing the timing action to repeat. Obviously, we can arrange to set the time delay to any value we require by using different values of C and R. It is usually convenient to drive the RC circuit and the voltage divider reference circuit with the same voltage as the op-amp positive power rail voltage, which in this case is +15 V.

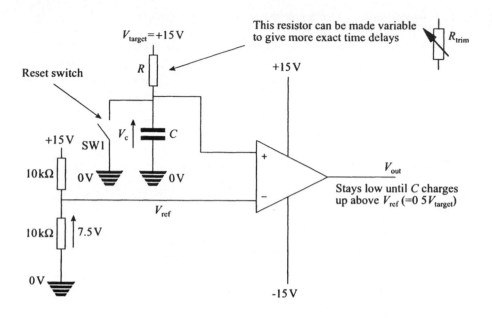

Figure 6.3

A simple way to set a known time delay is to recognize that an *RC* circuit *always* takes approximately $0.7CR$ seconds to reach half its target voltage ($0.5\ V_{\text{target}}$). This comes from the series *RC* charging equation

$$V_c = V_{\text{target}} \times [1 - \exp(-t/CR)]$$

(This equation tells us the value of V_c after any time t for specific C and R values. The CR product is called the 'time constant'.)

If we therefore make our time delay t equal to $0.7CR$ and set V_{ref} to $0.5\ V_{\text{target}}$, we have all the information we need to design the timer. Let us see an example of this.

EXAMPLE 6.3
Design a delay timer to switch an op-amp output to +14 V after 5 seconds.

SOLUTION
The circuit looks like Figure 6.3. We know that $t = 5$ s, therefore we set $0.7 \times CR = 5$ s, and hence

$$CR = 5/0.7 \text{ s}$$

We need to choose one of the C or R values as a starting point. Since real C values are more limited than real R values, we will start with a C.

Therefore, choose $C = 50\,\mu F$. Now we have

$$R = 5/(0.7 \times 50\,\mu F)$$
$$= 142.9\,k\Omega$$

Easy!

Although finding exact C and R values in real life may seem to be a problem, we can easily get round it by buying a variable resistor of about $270\,k\Omega$ and setting it somewhere near its half-way mark. We can always trim its value later by trial and error.

Self-assessment questions

SAQ6.3 Design a delay timer to give a delay of 45 s before an op-amp output switches from low to high.

SAQ6.4 If we used $V_{ref} \ll 0.5\,V_{target}$, would the delay time increase or decrease?

SAQ6.5 If we wanted very long (but not necessarily accurate) time delays, what would be a good value for V_{ref}?

SAQ6.6 If in Example 6 3 we replaced C by a value of $100\,\mu F$, what would the new delay interval become?

After a timed interval has elapsed, the capacitor charge is removed by closing the reset switch You will notice that there is no current-limiting resistance here to prevent a very large (short-lived) discharge current flow through the switch contacts. This current flow could easily result in permanent switch damage by welding the switch contacts together (arc-welding). To prevent this damage, a resistor would be adding in series with the switch to limit the current.

We have now looked at two types of simple but effective comparator circuit. We have seen how the circuit makes a comparison between an incoming voltage level and a reference voltage. By adding another op-amp we can produce a circuit which will compare an incoming voltage with *two* different reference voltage levels.

6.4 The window comparator

The window comparator is a new circuit which allows us to check if an incoming voltage level lies within a specified range of voltages. The specified range of voltages is called a 'voltage window'. Let us see the circuit for one Look at Figure

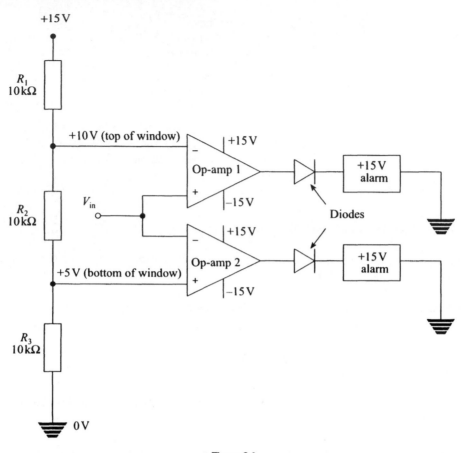

Figure 6.4

6.4. This particular example shows a two op-amp version of the window comparator You can see that there is a voltage divider circuit which fixes the voltage levels at two of the inputs. These voltages will set the top and bottom of the voltage window, as we will see shortly Notice that the input voltage is applied to two of the inputs at once.

We can now consider what will happen at the outputs of the op-amps as we apply different V_{in} values to the circuit

For op-amp 1, when V_{in} is above +10 V the non-inverting input is at a higher voltage than the inverting input. Therefore, the output goes high. With a little thought we can see that the output will be low whenever V_{in} is less than +10 V, and high whenever V_{in} is greater than +10 V. At the instant that V_{in} becomes fractionally higher than +10 V, op-amp 1 output goes high and sets off the alarm

For op-amp 2, when V_{in} is below +5 V the non-inverting input is at a higher voltage than the inverting input. Therefore, the output goes high. As for op-amp 1, reasoning leads us to the conclusion that the output will be high for all V_{in} values

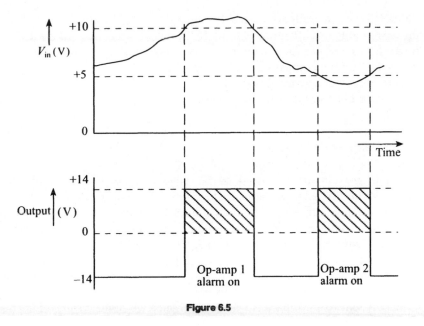

Figure 6.5

less than +5 V, and low for all values greater than +5 V. The alarm is set when the output goes high

The diodes in the circuit will allow current to flow in only one direction, i.e. from op-amp to alarm. They operate like a one-way valve, preventing current from flowing in the opposite direction when the op-amp outputs go negative This prevents damage to the alarm circuits when they are driven by the wrong polarity. Diode operation is covered in Chapter 7.

The window comparator is therefore a good circuit for detecting over-voltage or under-voltage conditions of V_{in}. The alarms are included to give you an idea of the types of application for which the circuit may be used. Figure 6.5 shows the V_{in} waveform which would set off each of the alarm circuits as it falls outside the specified voltage window.

We can control the window size quite easily by changing the values of the voltage divider circuit.

Self-assessment questions

SAQ6.7 Design a window comparator which will give an alarm when a voltage falls outside the +/−2 V range

SAQ6.8 Design a window comparator which will give an alarm when a voltage falls below 0 V and rises above +5 V.

SAQ6.9 If the V_{in} in Figure 6.4 were generated by a 15 V voltage divider circuit consisting of a thermistor and a 12 kΩ resistor as shown in part of Figure 6.2, the circuit could be used as an over- or under-temperature alarm. What range of thermistor values would cause:
1. the over-temperature alarm condition;
2. the under-temperature alarm condition?

6.5 The Schmitt trigger

The comparator circuits discussed so far have been switching circuits whose outputs have merely been responses to the input voltage conditions on the op-amp. We are now going to look at some comparators where the level of the output from the op-amp can influence what is happening to the input of the op-amp. This idea might seem familiar to you Remember negative feedback? All our early op-amp circuits had feedback connections from the output to the inverting input, providing a stabilizing negative feedback effect on them. Now we have negative feedback's destabilizing close relation, 'positive feedback'!

Generally, positive feedback is what we get when we connect some of the output to the non-inverting input If we look at Figure 6.6, we can see a comparator using some positive feedback. The main difference with this type of comparator is that the reference voltage is obtained from a voltage divider across the output of the op-

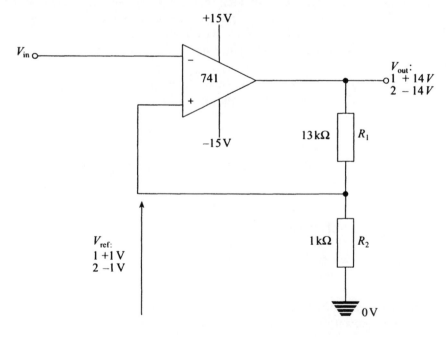

Figure 6.6

amp. It is obvious, therefore, that the size of the reference voltage will depend on the size of V_{out} from the op-amp. To analyse what happens, we will consider a specific V_{out} value.

We must first assume that V_{out} is sitting at its positive saturation level of about +14 V. Note that we have a V_{ref} of +1 V at the non-inverting input, produced by the voltage divider across the +14 V output and earth. This situation would be consistent with an applied V_{in} of anything less than +1 V Let us assume for the sake of argument that we have V_{in} = 0.9 V. Imagine now that we steadily increase V_{in} beyond +1 V towards, say, +2 V. The instant that V_{in} passes above +1 V by a few microvolts, the op-amp switches and V_{out} starts heading towards −14 V. As V_{out} heads towards −14 V, V_{ref} also drops and heads towards −1 V. This is a short-lived dynamic situation, where V_{ref} is actually increasing the difference input voltage to the op-amp, which in turn increases the rate at which V_{out} rushes towards its negative saturation value. Now we have a new situation where V_{ref} = −1 V. In the basic comparator (Section 6.1) the V_{out} switched back again when V_{in} crossed back over the same reference voltage. However, this is not the case here.

For the op-amp to switch back to V_{out} = +14 V, V_{in} now has to be reduced beyond the new reference voltage of −1 V. The waveform diagrams of Figure 6.7 demonstrate this action We can deduce, therefore, that the positive feedback

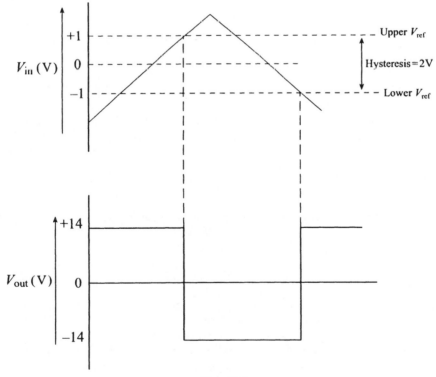

Figure 6.7

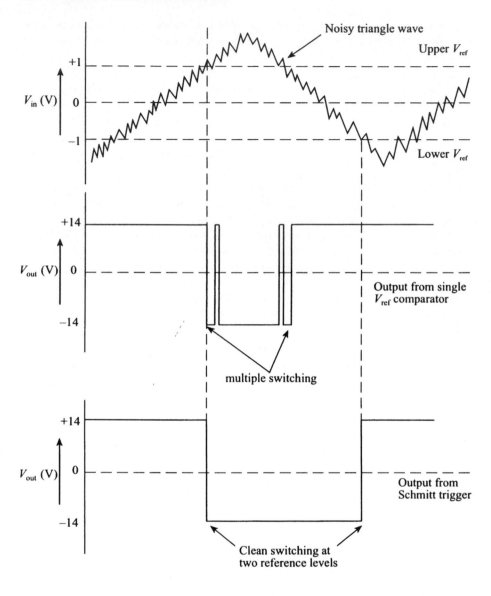

Figure 6.8

comparator has not one, but two, reference voltages. By altering the values of the resistors in the voltage divider we can increase or decrease the difference in the reference voltage values. The difference in the two values is known as 'hysteresis'. In the circuit in Figure 6.6 the hysteresis is [+1 V −(−1 V)] = 2 V. Comparator circuits like this are called 'Schmitt trigger' circuits and are commonly used to clean up noisy square wave signals.

Whenever a signal containing noise appears at the input of a comparator circuit, unpredictable and unwanted switching transients may be produced. Illustrated in Figure 6.8 we have a rather noisy-looking triangle wave which is to be used to switch some circuitry If this signal is used with a single V_{ref} comparator, the output contains unpredictable levels caused by the noise variations on the signal. If, however, it is input to a Schmitt trigger circuit with a 2 V hysteresis, the output ignores the noise variations and produces a clean square wave pulse This happens because it switches only when the signal *first* crosses the upper or lower reference voltage and is no longer responsive to local signal variations about that voltage.

Self-assessment questions

SAQ6.10 For the circuit shown in Figure 6 6, determine the new voltage reference values created if we replaced the resistors with the values $R_1 = 10\,k\Omega$ and $R_2 = 10\,k\Omega$

SAQ6.11 What would be the hysteresis of the new circuit in SAQ6.10?

SAQ6.12 Again, using the circuit of Figure 6 6, what would be the approximate values of the two reference voltages and the hysteresis if the power supply rails to the 741 op-amp were reduced to +/−9 V?

SAQ6.13 If we discovered that a signal contained more noise than we originally thought, would we arrange to increase the hysteresis value or decrease it?

6.6 The astable oscillator

Circuits for generating square waves are always of great use to the electronics designer. One of the easiest to build and use is the op-amp 'astable oscillator'. Before looking at the oscillator circuit itself, let us consider the name. Astable basically means non-stable. This circuit is considered astable because it will not flip into one condition and stay there. It is continually doing something which causes it to flip backwards and forwards from one condition to another. In other words, it continually oscillates between two semi-stable conditions The reason it is included here is that it is merely a combination of the delay timer circuit and the Schmitt trigger circuit Look at Figure 6.9 to see the two parts of the circuits combined to produce the oscillator.

The circuit works like this:

1. Assume initially that the capacitor C is uncharged and $V_{out} = +14$ V.
   Therefore, we can deduce that:

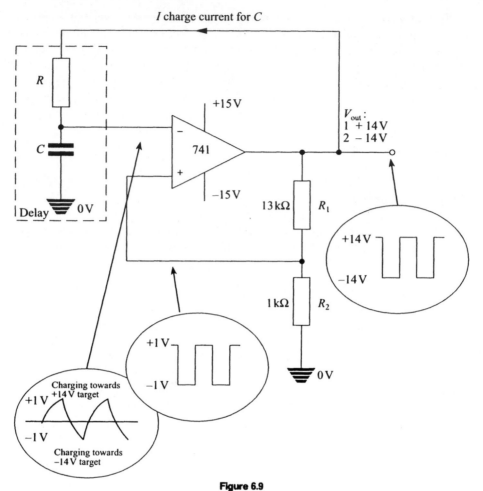

Figure 6.9

- the inverting input is at $0\,V$;
- the non-inverting input is at $+1\,V$ (by voltage divider action);
- this condition is consistent with our understanding of op-amp operation, i.e. if the non-inverting input is more positive than the inverting input, then the output will go to the positive saturation level.

The $+14\,V$ op-amp output can drive a charge current into the series CR circuit, causing the capacitor voltage to rise exponentially towards its target voltage of $+14\,V$. This, of course, is what happens but, when the capacitor voltage passes $+1\,V$, the op-amp will flip its output to $-14\,V$ because it now sees a lower voltage on its non-inverting input than on its inverting input.

2. For the next condition we have:
   - $V_{out} = -14\,V$;
   - slightly more than $+1\,V$ across C and on the inverting input;

- a new voltage of -1 V on the non-inverting input (by voltage divider action).

This condition causes the op-amp output to charge the series CR circuit towards its new -14 V value The capacitor voltage therefore drops exponentially from $+1$ V towards -14 V. But when the capacitor voltage passes -1 V, the op-amp will flip its output to $+14$ V because it now sees a higher voltage on its non-inverting input than on its inverting input.

   This takes us back to the previous condition, and so on.

   The circuit flips backwards and forwards at a rate governed by the charge and discharge times of the CR circuit. The circuit gives us a square wave on V_{out} and across R_2, and a waveform resembling a triangle wave across C, as shown in Figure 6 9. We can easily see that the V_{out} period is dependent on the sizes of C and R. If we analysed the time it takes a capacitor to charge and discharge to the appropriate reference voltages appearing on the non-inverting input, we would produce the equation for the period (T) of V_{out}, which is

$$T = 2CR \, \log_e(1 + 2R_2/R_1)$$

The frequency (f) is therefore given by

$$f = [2CR \, \log_e(1 + 2R_2/R_1)]^{-1}$$

EXAMPLE 6.4

What is the output frequency of the circuit in Figure 6 9 if $C = 1\,\mu\text{F}$ and $R = 100\,\text{k}\Omega$?

SOLUTION

We will use the expression

$$f = [2CR \, \log_e(1 + 2R_2/R_1)]^{-1}$$

First we will find the values of CR and R_2/R_1

$$CR = 1 \times 10^{-6} \times 100 \times 10^3$$

$$= 0.1 \text{ s}$$

$$R_2/R_1 = 1 \text{ k}\Omega/13 \text{ k}\Omega$$

$$\approx 0.077$$

Now

$$2R_2/R_1 \approx 0.154$$

$$\log_e(1 + 2R_2/R_1) \approx \log_e(1.154)$$
$$\approx 0.143$$

Therefore

$$f = (2 \times 0.1 \times 0.143)^{-1}$$
$$= (0.0286)^{-1}$$

giving

$$f \approx 34.9$$
$$\approx 35 \text{ Hz}$$

This is quite a low frequency oscillator.

In Example 6.4 we used different values of R_1 and R_2, but very often it is easy (and convenient) to use $R_1 = R_2$. The advantage of this is that the output frequency equation becomes easier to remember! The original equation

$$f = [2CR \log_e(1 + 2R_2/R_1)]^{-1}$$

becomes

$$f = [2CR \log_e(3)]^{-1}$$

Therefore

$$f \approx (2.2CR)^{-1}$$

This is an expression that we could memorize

A circuit designer often looks for the easiest way to verify an idea roughly. The fine tuning can come later. The expression $f \approx (2.2CR)^{-1}$ could be approximated even further to $f \approx 1/(2CR)$. You must agree that we cannot hope for a simpler expression than this. Remember that this is for $R_1 = R_2$ only.

Self-assessment questions

SAQ6.14 We have built an op-amp astable oscillator with $R_1 = R_2 = 10 \text{ k}\Omega$,

$C = 0\ 01\ \mu F$ and $R = 1\ k\Omega$. Find the approximate and more exact values of the resulting output frequency.

SAQ6.15 Without doing the calculations, what would be the effect on frequency if we:
1. increased R?
2. decreased C?
3. increased the power supply to the op-amp?
4. reduced R_2?
5. reduced R_1?

SAQ6.16 Are the two square waves from V_{out} and across R_2 in phase or out of phase?

SAQ6.17 For $C = 100\ nF$, $R = 5\ k\Omega$, $R_1 = 10\ k\Omega$, $R_2 = 40\ k\Omega$ and supply rails to a 741 op-amp astable circuit of $+/-10\ V$, find T, f and the size of the waveforms at V_{out} and across R_2.

SAQ6.18 If the purpose of building an op-amp astable circuit is to utilize the (almost) triangle wave output across C, which of the following options would provide a better shaped triangle wave (size does not matter): large CR, small CR, large power supply, small power supply, large R_2 or small R_2?

SAQ6.19 Design an op-amp astable circuit to give a 5 kHz square wave output.

SAQ6.20 What would be the advantage in replacing the value of R by a variable resistance?

Checklist

Can you now confidently design:
- a circuit which will detect a change of voltage or a change of resistance?
- a circuit which will ring an alarm if the temperature of a room gets too high?
- a circuit which will detect if a voltage is within a specified range of values?
- a circuit which will ring an alarm if the temperature of a room gets either too high or too low?
- a circuit to ring an alarm after a specified time delay?
- a circuit which will clean up a square wave containing lots of noise?
- a circuit which will produce a square wave at a specified frequency?
- a circuit which will produce a waveform resembling a triangle wave at a specified frequency?

Tutorial questions

1   Using a typical 741 value for A_{vol}, find the input voltage difference required to produce $V_{out} = 2\,V$.

2   Design a circuit to switch on a 9 V (5 mA) radio when the resistance of a light-dependent resistor falls slightly below 5 kΩ. (The 741 output can deliver up to 10 mA) You may assume that the radio is protected against a negative power supply

3   Redesign the circuit in question 2 so that the radio would turn off instead of on.

4   Using the RC charging equation, prove that an RC circuit takes $0.7CR$ s to reach approximately half its target voltage.

5   Design a delay timer to switch an op-amp output from $-9\,V$ to $+9\,V$ after 35 s.

6   Redesign the circuit in question 5 so that the output switches from high to low after 35 s.

7   Can you think of reasons why very long time delays are less accurate using this type of delay timer circuit?

8   Design a delay timer to switch an op-amp output from $-5\,V$ to $+5\,V$ after 800 ms.

9   Design a comparator circuit to switch on an alarm if a voltage falls outside the ranges $+/-10\,V$, 0 to $+1\,V$ and 0 to $-3\,V$.

10   Unity gain inverting amplifiers can often be used to invert the polarities of switching voltages. With this in mind, see if you can design a comparator circuit to switch on an alarm if a voltage falls inside the ranges $+/-2\,V$, 0 to $+8\,V$ and $+5$ to $+6\,V$.

11   If the hysteresis of a Schmitt trigger were reduced, would the circuit's ability to ignore noisy signals be improved?

12   Design a Schmitt trigger with a hysteresis of $+5\,V$.

13   If the hysteresis of a Schmitt trigger is set at $+2\,V$ and the power supply to the circuit is reduced to 75% of its original value, what is the new hysteresis value?

14   What type of feedback is used in the Schmitt trigger circuit?

15   A $5\,V_p$ waveform used as input to a Schmitt trigger contains 200 mV of noise. What would be the most useful hysteresis value: 150 mV, 300 mV, 2.5 V or 5 V?

16   With reference to the Schmitt trigger circuit in Figure 6.6, what would be the hysteresis if $R_1 = 180\,kΩ$ and $R_2 = 10\,kΩ$?

17   With reference to the astable oscillator circuit in Figure 6.9, assume $CR = 5\,s$ and determine the output frequency if $R_2 = 10R_1$.

18   Draw fully labelled waveform diagrams of the op-amp inputs and output in question 17.

19   Design an astable oscillator circuit with an output frequency of 1 5 kHz

20 Explain why it is a good idea to include a resistor in series with the reset switch in the delay timer circuit.

Further reading

Bogart T F (1986) *Electronic devices and circuits*, section 14 7 Merrill, Columbus, Ohio

Horowitz P & Hill W (1987) *The art of electronics*, sections 3 22–3 Cambridge University Press, Cambridge, England

CHAPTER 7

Diodes and their applications

Aims and objectives

This chapter will introduce you to the standard diode, the light-emitting diode (LED) and the zener diode. It will explain what they do and how they can be used in different circuits to perform many types of operation on waveforms. You will be shown different ways of modelling diodes to determine circuit action.

The chapter will also give you all the information required to specify, design and build a power supply using zener or integrated circuit voltage regulation.

After studying this chapter you will be able to:

- *explain what a diode can do*
- *interpret the diode characteristic curve*
- *interpret the diode equation*
- *understand the effects of temperature variations on the diode*
- *recognize the difference between static, average and dynamic diode resistance values*
- *recognize the use of a simple diode model in circuit analysis*
- *design and use light-emitting diodes (LEDs)*
- *design simple voltage clipping circuits*
- *recognize the difference between half-wave and full-wave rectifiers*
- *fully design a power supply circuit*
- *design and use simple voltage regulators with zener diodes*
- *recognize why zener diodes are generally useful in creating voltage references*
- *design circuits with fixed voltage and variable voltage integrated circuit regulators*

7.1 Introduction to the diode

A simple introduction to the diode would be to describe it merely as a one-way valve for electric current. It will allow current to flow in one direction but not in the other. In this chapter we will investigate a number of diode circuits where the circuit operation is no more complicated than this.

As we delve a little further into diode operation we will find that there are one or two other things that will describe the diode operation with more accuracy. Diodes are easy to understand and very useful devices to have around Their beauty is in their simplicity, and creative designers will always be finding novel areas where they can be applied. In this chapter we are not going to introduce any semiconductor theory to explain why the diode works: we are merely going to investigate what it will do and how to use it.

Diodes are manufactured from many different types of semiconducting material such as silicon (Si), germanium (Ge) and gallium arsenide (GaAs). Each material has its own characteristics, some of which we will mention briefly. However, for most of this chapter we will consider the standard diode to be made from silicon So unless a diode material is specified, we will assume it to be silicon

7.2 The symbol and the real thing

The diode is a two terminal device with a symbol as shown in Figure 7.1. We need to know which way round to connect it, therefore, one end is called the 'cathode' (k) and the other the 'anode' (a). The diode is made from two types of semiconductor material the anode from P-type and the cathode from N-type. Because of the types of semiconductor used, the diode is often referred to as a 'PN junction diode'.

We have already stated that the diode allows current flow in one direction only. The symbol rather sensibly contains an arrow which indicates which way the current flows The cathode end is a bar, to remind us that current will not enter at this end. When we buy a real diode we find that diodes come in all shapes and

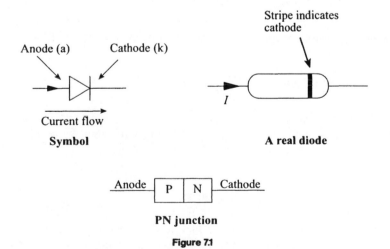

Stripe indicates cathode

Anode (a)   Cathode (k)

Current flow

Symbol

I

A real diode

Anode | P | N | Cathode

PN junction

Figure 7.1

sizes, but they all have one thing in common. the mark on the device to indicate the cathode. Figure 7.1 illustrates the most common shape of a real diode that you might see, but there are many others!

7.3 Diode characteristics

If we experimented with the diode we would find that its behaviour can be described by four simple observations

1. The diode conducts easily in one direction but not very well in the other
2   To make a diode conduct we must create a potential difference or voltage across the device, with the anode being more positive than the cathode.
3. The size of the voltage required across the diode for conduction is a characteristic of the diode material For example, it is 0.6 V for a silicon diode.
4. We obtain a massive increase in the current flow through the diode when we increase the voltage across its ends by only a few millivolts.

These four observations indicate the diode operation. We now need to learn some of the general terms used in diode work.

7.4 Forward bias

When a voltage is applied across the diode, making its anode more positive than its cathode, it will conduct easily. This is called 'forward biasing the diode'.

7.5 Reverse bias

When a voltage is applied across the diode, making the cathode more positive than the anode, it will not conduct. This is called 'reverse biasing the diode'.

7.6 Threshold voltage

The size of the voltage required to forward bias a diode is called the 'threshold' or 'barrier' voltage (V_T). For silicon, $V_T \approx 0.6$ V; for germanium, $V_T \approx 0 3$ V, and for gallium arsenide, $V_T \approx 1 6$ V

7.7 Reverse saturation current

In reverse bias a diode will hardly conduct at all There is, however, a small but measurable reverse current flow called the 'reverse saturation current' (I_s)

7.8 The diode characteristic curve

As in any other engineering discipline, the quickest way to demonstrate what a device will do is to draw a picture. The best picture to draw here is the graph which relates the diode current flow to the applied diode voltage in forward and reverse bias conditions This is called the 'diode characteristic curve'. This curve is shown in Figure 7 2.

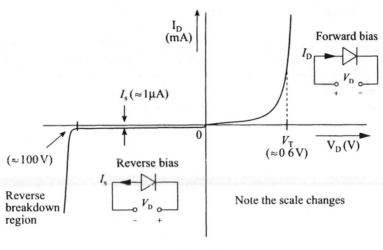

Figure 7.2

There are a number of observations that can be made from Figure 7.2 We can see that as the forward bias voltage is steadily increased from zero towards V_T the current starts to rise, but most of the rise occurs after V_T. In fact, this rise is exponential, but we often treat the rise in current above the V_T value as being almost linear. In the reverse bias region we can see that very little current (I_s) flows. As expected, the actual amount will vary between diodes, but normally Ge diodes have I_s of a few microamps and, for Si diodes, $I_s \approx 1 \mu A$, although Si diodes can be purchased which have a much lower value of I_s than this If we increase the reverse bias voltage to very large values, we reach a breakdown region where a large reverse current flows. For standard diodes this voltage is usually high (≈ 100 V) and is a region we avoid using. The data sheets for real diodes will normally specify this reverse breakdown voltage as either peak inverse voltage (PIV) or reverse voltage (V_R)

7.9 The diode equation

We already known that the I_D rise is exponential with V_D If we were to express this mathematically, we would use the diode equation which tells us

$$I_D \approx I_s \times \exp(V_D/nK)$$

where $K = 0.026\,V$ and is a constant at room temperature, $n = 1$ for Ge and $n = 2$ for Si up to V_T, and $n = 1$ beyond V_T.

It is not important to memorize this equation: merely to note that it is a mathematical way of representing the diode characteristic curve

7.10 Limiting the size of the diode current

From Figure 7.2 or from the diode equation, you can see that a small increase in V_D will cause very large currents to flow. There is a major problem here concerning the fact that increased current flow causes the diode to heat up. Too much current and too much heat mean that there is not much chance of the diode lasting very long. In real circuits the maximum current flow *must* be limited to prevent diode damage by inserting an extra external resistance, called a 'current-limiting resistor', in the diode circuit. Figure 7.3 shows the circuit we would use.

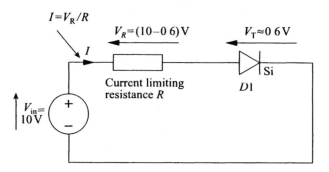

Current is limited by the voltage across R

Figure 7.3

EXAMPLE 7.1
For the circuit in Figure 7.3, what value of current-limiting resistance would be used to ensure that the diode current never exceeded $10\,mA$?

SOLUTION
Using

$$I = V_R/R$$

we have

$$10\,mA = (10 - 0.6)V/R$$

Therefore

$$R = 9.4 \text{ V}/10 \text{ mA}$$

$$= 0.94 \text{ k}\Omega$$

7.11 Effect of temperature

Unfortunately, the diode current flow is rather temperature-dependent, but it is reasonably predictable. The two main things to remember are.

1   As temperature increases, V_T decreases In fact, for every 1 °C rise we get a 2 mV fall in V_T.
2.   As temperature increases, I_s increases For every 10 °C rise we get a doubling of the I_s value.

Let us see a picture of this to help remember the effect; see Figure 7.4.

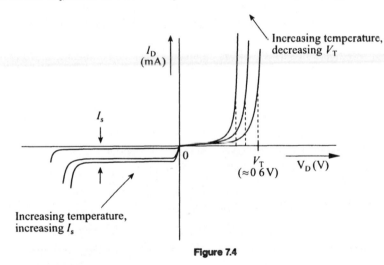

Figure 7.4

EXAMPLE 7.2
If a diode whose $V_T = 0$ 6 V and $I_s = 2 \mu\text{A}$ suffers a temperature rise (dT) of 20 °C, what happens to V_T and I_s?

SOLUTION
V_T drops to V_{T1} and I_s rises to I_{s1}, where

$$V_{T1} = 600 \text{ mV} - (20 \times 2) \text{ mV}$$

$$= 560 \text{ mV}$$

$$I_{s1} = (2\mu\text{A} \times 2^d)$$

where d is the number of $10\,°C$ intervals in dT

$$d = dT/10$$
$$= 20/10$$
$$= 2$$

Therefore

$$I_{s1} = 2 \times 2^2$$
$$= 8\mu A$$

7.12 Resistance of the diode

We have already mentioned that we often assume the rise in current above V_T to be linear (although we know it is exponential). It also tends to make the diode behave as if it had a resistance in series with its electronic valve mechanism. If we could find a value for this resistance we would be able to model the diode behaviour in a circuit quite easily. This section demonstrates what these resistances are and how they affect the diode circuitry It should be stressed, however, that consideration of the diode resistance in a circuit should be your second choice of analysis of a diode circuit. The first choice should be to treat the diode merely as a one-way valve with a voltage drop across it. Once you have understood the diode circuit you can then investigate the effect the diode resistance might have on it.

7.12.1 D.c. resistance (static resistance)

First look at Figure 7 5, where three points labelled 1, 2 and 3 have been shown on a typical diode curve. If we use Ohm's law to find the d c. diode resistance R_D at point 1, we obtain

$$R_D = V_D/I_D$$
$$= 0.7 \text{ V}/25 \text{ mA}$$
$$= 28\,\Omega$$

At point 2

$$R_D = 0.4 \text{ V}/2 \text{ mA}$$
$$= 200\,\Omega$$

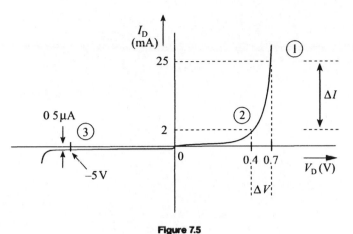

Figure 7.5

and at point 3

$$R_D = -5 \text{ V}/ - 0\ 5\mu\text{A}$$

$$= 10 \text{ M}\Omega$$

It is apparent from this that the resistance to current flow in the reverse direction is extremely high, which is just what we would expect The forward resistance is rather low and decreases with the current flow. This information tells us that when we use a diode in a circuit it will behave as a one-way valve with a small resistance. If we know the value for R_D, we can use it to model our diode in a circuit to obtain more accurate values of the circuit current If, and when, we apply a model to analyse a diode circuit, it is quite acceptable to model the reverse bias resistance merely as an open switch and to assume that no current flows in the circuit

7.12.2 Average diode resistance

Because the static resistance differs widely from small and large current flows, it is more usual to use an average resistance value (R_{av}) over the range of currents. To do this we merely take two extreme points on the forward bias part of the diode curve (points 1 and 2 in Figure 7.5) and use $R_{av} = \Delta V/\Delta I$, where ΔV is the difference in voltage and ΔI is the difference in current between the points. Using Figure 7.5 we have an average resistance

$$R_{av} = (0.7 - 0.4) \text{ V}/(25 - 2) \text{ mA}$$

$$= 13\ \Omega$$

Normally, of course, when we find the gradient of a curve we find $\Delta I/\Delta V$, which gives us $1/R_{av}$ Therefore, the reciprocal of the gradient gives us the diode

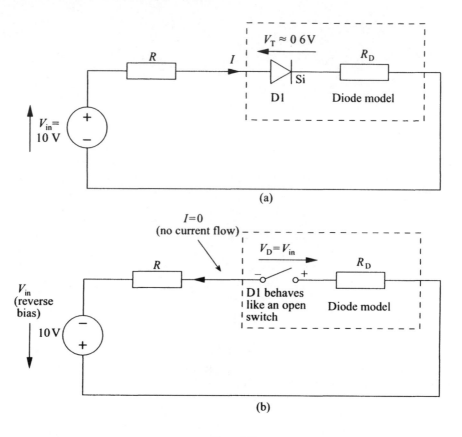

Figure 7.6

resistance R_{av}. The value for the average resistance is dependent on how much the diode curve leans outwards. The further it leans, the higher the R_{av} value. In other words, the gradient of the diode curve above V_T is very important.

A diode model can be drawn whenever we have a real diode in a circuit Using this idea we could redraw the circuit in Figure 7.3 to include the diode resistance This is done in Figure 7.6(a) The difference that any diode resistance makes is to reduce the overall current a little. An example easily illustrates this.

EXAMPLE 7.3

What is the diode current in Figure 7 6(a) if the diode resistance is $20\,\Omega$ and a current limiting resistor of $1\,k\Omega$ is used?

SOLUTION

Note that the voltage supply is d c , therefore we use R_D. We need to remember that the diode action takes up 0.6 V, effectively reducing the supply voltage to the rest of the components in the circuit. We can therefore see that this reduced voltage (V_{circuit}) is

$$V_{\text{circuit}} = (V_{\text{in}} - 0.6)$$

$$= (10 - 0.6)$$

$$= 9.4 \text{ V}$$

The circuit current (I) is therefore

$$I = V_{\text{circuit}}/(R + R_{\text{D}})$$

$$= 9.4 \text{ V}/1020\,\Omega$$

$$= 9.2 \text{ mA}$$

You may ask, what happens to this model during reverse bias? If we apply a reverse bias voltage at V_{in}, virtually no current flows and it is easiest to treat D1 as an open switch, as shown in Figure 7.6(b). This technique holds for all diode circuit analysis unless, of course, we are specifically interested in the flow of I_{s}.

7.12.3 Dynamic resistance

When we use the diode with only small a.c. signals, we find that ΔV and ΔI are also very small. This means that we are interested in the change of voltage and current around a small region of the graph. Finding the slope $\Delta V/\Delta I$ for very small values involves a bit of calculus: we differentiate the relationship between I and V, and then find its reciprocal. The relationship is, of course, the diode equation. You can try this differentiation in one of the SAQs a little later. Assuming the mathematics has been done, we obtain a simple expression for dynamic resistance (r_{d}), given by

$$r_{\text{d}} = (0.026/I)\Omega$$

where I is the steady d c. current flow through the diode at room temperature.

This dynamic resistance is of use only when the ΔI and ΔV variations are occurring somewhere in the conducting part of the curve Look at Figure 7.7 to see this.

The dynamic resistance is also sometimes called the 'a c. resistance'. Typically, the value of r_{d} is small compared with R_{av} or R_{D}.

EXAMPLE 7.4
Determine the value of r_{d} at point 1 in Figure 7 5 and compare it with that of R_{D} at the same point.

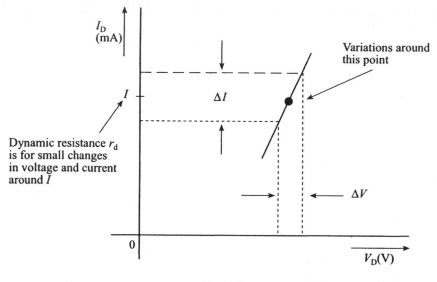

Figure 7.7

SOLUTION

Previously, we found at point 1 $R_D = 28\,\Omega$. At this point we have a steady current flow $I = 25\,\text{mA}$. Therefore

$$r_d = 0.026/25\ \text{mA}$$

$$= 0.026/0.025$$

$$\approx 1\,\Omega$$

We note that r_d is much less than R_D, as expected.

7.12.4   Which diode resistance do we use?

To analyse a diode circuit, follow the four steps:

1.   First, ignore any diode resistances until you are happy with how the circuit works.
2.   Without doing any calculations, try to determine whether a resistance of a few tens of ohms would have any more than a negligible effect on the circuit. If not, do not worry about diode resistance at all (Hint· if all the other circuit resistances are $500\,\Omega$ or more, it is unlikely that you need to bother with diode resistance!)
3.   If you decide that diode resistance might be of importance, you must include a diode resistance value in your analysis. (See the next step.)

4. Use R_{av} if the circuit is being driven by a large a.c. voltage signal (anything more than about 1 V). This is by far the most common type of application. R_{av} can be estimated from the diode curve. Use r_d if the circuit is being driven by a small a.c. voltage signal with a d.c. offset of about V_T. $r_d = 0.026/I$, where I is the forward bias current caused by the offset voltage. Use R_D if the circuit is driven by only a d.c. voltage supply. R_D can be approximated from the diode curve.

Self-assessment questions

SAQ7.1 Using Figure 7.6(a), what would be the diode current if a germanium diode were used in place of the silicon diode, given $R = 10\,k\Omega$ and $R_D = 40\,\Omega$?

SAQ7.2 Using Figure 7.6(a), what current limiting resistance would be used to limit the silicon diode current to 50 mA with $V_{in} = 5\,V$ and $R_D = 50\,\Omega$?

SAQ7.3 If the diode in a circuit containing a germanium diode were replaced by a silicon diode, would the circuit current rise or fall?

SAQ7.4 If $I_s = 100\,nA$ for a diode at 20 °C, what value would the I_s of the diode reach if the temperature dropped to 0 °C?

SAQ7.5 If $V_T = 0.3\,V$ for a diode at 80 °C, what value would the V_T of the diode reach if the temperature dropped to 10 °C?

SAQ7.6 Why do we need to take the *reciprocal* of the gradient of the diode curve to determine the diode resistance?

SAQ7.7 What happens to the value of R_D when the diode current increases?

SAQ7.8 What is the dynamic resistance value of a diode carrying a steady current of 1 mA at room temperature?

SAQ7.9 Differentiate the diode equation with respect to V_D to prove that $1/r_d = I/0.026$, and hence $r_d = (0.026/I)\,\Omega$.

SAQ7.10 If a voltmeter placed across the diode in Figure 7 3 gave a reading of 10 V, what would you diagnose the fault to be?

7.13 Using light-emitting diodes (LEDs)

An LED is a special type of diode made from gallium arsenide (GaAs) or gallium phosphide (GaP) which emits radiation when it is forward biased. The most commonly used LEDs emit visible radiation in the form of green, red or yellow light, and are used as indicators for equipment. Other common types emit infra-red radiation (just beyond the red end of the visible spectrum) and are used for communications and the remote control of equipment. (Remember that trusty TV remote controller which means you do not have to move from your armchair all night? It is all thanks to the LED) The LED symbol indicates the radiation being emitted and a marking on the component distinguishes the anode from the cathode (see Figure 7.8).

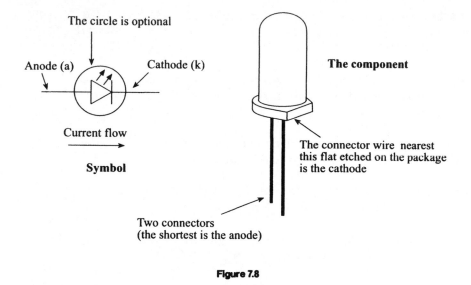

Figure 7.8

The way we use and analyse LEDs in circuits is similar to that for any other diode. The two main differences are that the V_T is larger than most other diodes and that LEDs can be damaged by relatively small reverse bias voltages

7.13.1 Light emission

The amount of light emitted by the LED is directly proportional to the forward current flow through it. Therefore, for a bright LED go for a higher I value. One word of warning, however: the brighter the LED, the shorter life span it will have. This is common sense since any component driven to its maximum performance will burn out more quickly It is best to aim for a happy medium.

7.13.2 Change of V_T

The V_T of an LED depends on the wavelength of light emitted: red LEDs which emit longer wavelength light waves will have slightly smaller V_T values than green LEDs. Typically, V_T will be 2.2 V for a green LED, 1.6 V for a red LED and 1.2 V for an infra-red LED If in doubt use $V_T \approx 1.8$ V. This is a reasonable general value.

7.13.3 LEDs in action

Let us look at an example of LEDs in circuits

EXAMPLE 7.5

Design a circuit to drive an LED from a 12 V supply. The LED has $I_{max} = 30$ mA.

SOLUTION

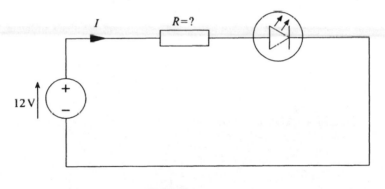

Figure 7.9

The circuit is shown in Figure 7.9. Using $V_T = 1.8$ V, the voltage V_R across R is $12 - 1.8 = 10.2$ V We choose I to be slightly less than the I_{max} value, say $I = 20$ mA. Therefore

$$R = V_R/I$$
$$= 10.2 \text{ V}/20 \text{ mA}$$
$$= 0.51 \text{ k}\Omega$$
$$= 510 \Omega$$

For suitable operation, any value of $R > 510 \ \Omega$ is fine.

Self-assessment questions

SAQ7.11 What happens to the brightness of the LED if we choose 1 kΩ rather than 510 Ω?

SAQ7.12 Why do we not use $I = I_{max}$?

Example 7.5 uses an LED in a d c. circuit. If, however, we use an a c. supply, we must be careful of the reverse voltage. Remember LEDs do not like large reverse voltages. Look at Figure 7.10. When the a c voltage peaks in the positive half-

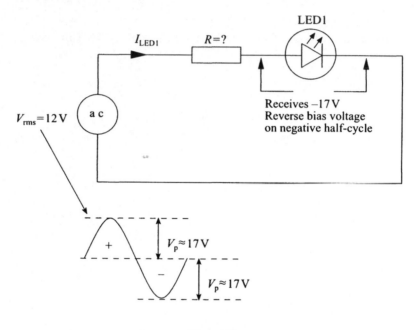

Figure 7.10

cycle, LED1 is forward biased and current flows clockwise around the circuit, its size being limited by R. When the a.c. voltage peaks in the negative half-cycle, LED1 is reverse biased, no current flows and the full V_p voltage is applied across the reverse biased LED. (Remember the open-switch reverse bias model?) This could damage the LED beyond repair! The solution to this problem is to limit the reverse bias voltage across the LED by placing another diode in parallel with it but the other way round In this way the extra diode becomes forward biased during the negative half-cycle, causing only the diode's V_T to appear across the reverse biased LED1. This technique is shown in Figure 7.11, where the protecting diode is another LED, called LED2. The extra advantage of this is that you get light during both half-cycles. More light, higher efficiency!

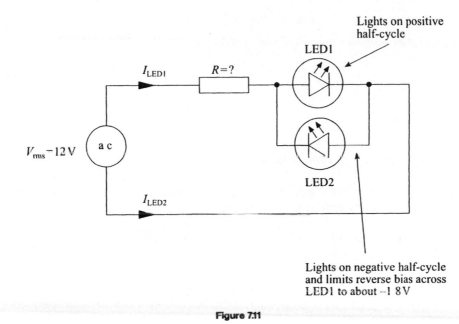

Figure 7.11

The interesting point to note about this technique is that each LED protects the other against high reverse bias voltages. The protecting diode added in the circuit can just as easily be a standard diode but, of course, we get no visible light from the circuit during the negative half-cycle

EXAMPLE 7.6
We want to use an LED as a power-on indicator across the UK mains supply. Design a suitable circuit if $I_{max} = 50\,mA$ for the LED

SOLUTION
To protect the LED against the very high reverse bias, we must use a protecting diode D1. We will choose a standard Si diode for this With reference to Figure 7.11, LED2 will be replaced by a Si diode and $V_{rms} = 230\,V$. (We must note that mains voltage is $230V_{rms}$ and has a tolerance of $+10\%/-6\%$, as defined by the European Committee for Electrotechnical Standardisation.)

We must work out the V_p value and the size of R. First we will choose $I < I_{max}$, and therefore we will take $I = 40\,mA$. The maximum $V_{rms} = 230$ ($+10\%$) $= 253V_{rms}$. Therefore

$$V_p = 253 \times 1.414$$

$$\approx 358 \text{ V}$$

Allowing for $V_T = 1.8$ across the LED, we have

$$R = (358 - 1.8) \text{ V}/40 \text{ mA}$$
$$= 8.9 \text{ k}\Omega$$

We would choose the next available resistor value up from this.

Self-assessment questions

SAQ7.13 In Example 7 6 what is the maximum reverse bias voltage that will appear directly across D1 and the LED?

SAQ7.14 In Example 7.6 what will be the current flow in the circuit during the negative half-cycle?

SAQ7.15 With reference to Example 7.6, diode D1 needs to have an I_{max} of what value? (Hint: remember not to run diodes to their full specification)

SAQ7.16 Here is a useful design problem. Given two LEDs, one red and one green, each with $I_{max} = 30$ mA, design a circuit to indicate when a 9 V battery has been connected correctly or incorrectly (i.e. the wrong way round) to a circuit.

7.14 Diode voltage-clipping circuits

We now know that diodes have a characteristic voltage drop across them when they are conducting. This diode property is of great use in voltage-clipper circuits. So what is a voltage clipper? A circuit which prevents a voltage level from exceeding a certain maximum or minimum value is known as a 'voltage clipper'. Voltage clippers are used to protect sensitive circuits from receiving unwelcome and potentially damaging levels of voltage.

7.14.1 Positive voltage clippers

Let us look at a simple voltage clipper. In Figure 7.12 we can see a circuit identical to that shown in Figure 7.3 except that it has been redrawn slightly and has output terminals attached. This circuit now has a purpose other than an academic exercise in simple diode theory! What is that purpose? Well, we can see that $V_{out} \approx 0\ 6$ V, so what will happen if we increase V_{in} to, say, 20 V? The diode current will increase and still the new $V_{out} \approx 0.6$ V. Strictly, we would find that V_{out} increases very slightly because of the effect of R_D, but not enough to worry about. This circuit

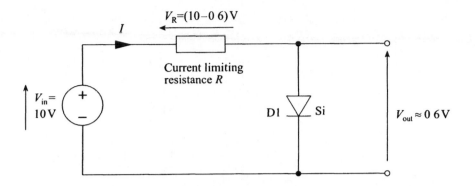

Diode current is limited by the voltage across R

Figure 7.12

therefore prevents any voltage larger than about 0.6 V from reaching V_{out}. In other words, it 'clips' voltages that are bigger than 0.6 V. Hence it is called a 'voltage clipper'.

This circuit is of more use if it is used with an a.c. input. Look at Figure 7.13 in which we have an input signal with positive and negative portions. A brief analysis of the circuit tells us that:

1. Whenever V_{in} is more than about 0.6 V, D1 conducts (or switches on) and limits V_{out} to 0.6 V.
2. Whenever V_{in} is less than 0.6 V or negative, the diode is off, no current flows and the full V_{in} voltage is dropped across D1 and therefore across V_{out} also.

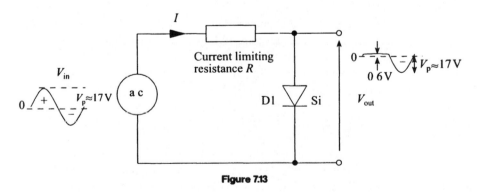

Figure 7.13

A simple modification to this circuit produces a more useful variable clipping voltage level. The idea is to raise the voltage on the cathode by a few volts to make it positive. In Figure 7.14 this voltage is labelled V_{clip} and it is shown as being

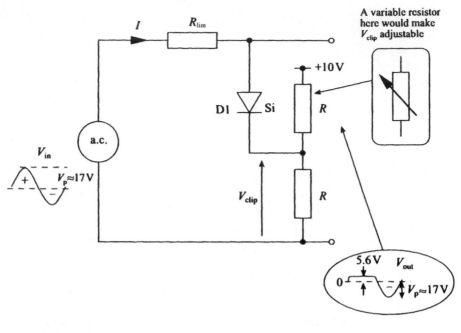

Figure 7.14

produced by the now familiar voltage divider circuit. The voltage required on the anode to make the diode conduct is now $(V_{clip} + V_T)$. You can easily see that if $V_{clip} = +5\,V$, the new clipping voltage will be $(5\,V + 0.6)$ and V_{out} will never rise above $5.6\,V$. Obviously, the size of the clipping voltage can be changed by the designer by the insertion of a variable resistor in place of one of the voltage divider resistances

In the variable voltage clipper circuit in Figure 7.14 it is important to note that the voltage divider circuit must be designed to be a good stiff voltage source which will not vary its V_{clip} voltage when diode current flows through it. The way to achieve this is always to make the voltage divider resistors small compared with the current limiter resistor R_{lim}.

7.14.2 Negative voltage clippers

Clippers for negative voltages are easily produced by connecting the cathode of the diode to the output. The circuit in Figure 7.15 clips anything below about $0.6\,V$. In the same way as with the previous clipping circuit, we can vary the clipping voltage by providing a different voltage on the anode of the diode. For example, if the circuit in Figure 7.15 were arranged so that a fixed $-5\,V$ source was connected to the anode, the circuit would be modified to prevent V_{out} from dropping below $-5.6\,V$

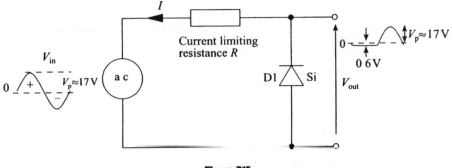

Figure 7.15

Two final notes before leaving voltage clippers:

1. It is perfectly acceptable to alter the clipping potential by using lots of diodes in series instead of extra voltage supplies. See Figure 7.16 for the idea Every extra diode added in series increases the voltage clipping potential by V_T In the circuit shown, silicon diodes are used to give a 3 V clipper.
2. These sorts of circuit are not usually used with pure sine wave signals; more often, they form a protective limiting action between a signal from a transducer (such as a microphone or a light sensor) and subsequent circuitry (perhaps an analog to digital converter). This idea is also shown in Figure 7.16.

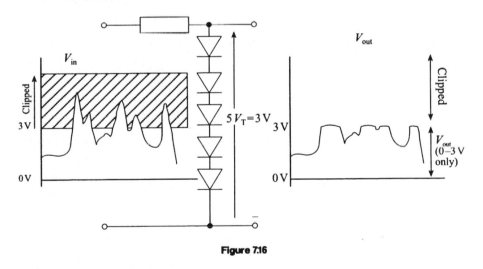

Figure 7.16

7.15 Diodes in power supplies

The most obvious place where you see diodes in use is in the old favourite power supply unit. The purpose of a power supply unit is to convert the mains voltage from the incoming electricity supply into a constant voltage suitable for driving electronic circuits. In doing this conversion, two major aspects of the mains voltage must be tackled:

1.   The mains voltage is a $230V_{rms}$ sine wave with a zero d.c. offset Since this voltage alternates between being positive and negative, it also causes current to alternate in any load circuit connected to it. For this reason it is called an 'alternating current source' or an 'a.c. source' or, more simply, just 'a.c.' To convert this into a useful form for a circuit, it must be converted into a direct (or constant) current source, generally known as a 'd c. source' The circuit performing this conversion is called a 'rectifier' and we will look at three types of rectifier circuit later.

2.   The mains voltage is nearly always too large to use directly in any circuit containing electronics. In the UK, the mains voltage is rated at about $230V_{rms}$ and, since most electronics circuitry works at voltages below 50 V, the size of this voltage must therefore be reduced. The component used to reduce (or transform) this voltage is the transformer and we will look at this first.

The main modules of the power supply can be illustrated by the diagram in Figure 7.17 and we will look at each of them separately.

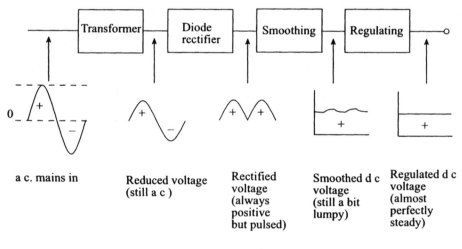

Figure 7.17

7.15.1 The transformer

In Chapter 1 we discussed briefly the effect of moving a magnet through an inductor coil to cause an induced voltage across the inductor. This connection between electric and magnetic fields is called 'electromagnetism' and works both ways. The electromagnetic effect also causes a magnetic field to be produced in the

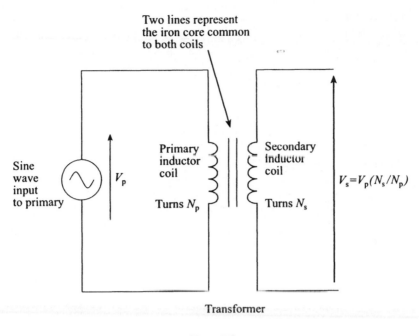

Transformer

Figure 7.18

coil when a varying external voltage is applied across the ends of the inductor. This is the basis of the transformer.

Referring to Figure 7.18, if two inductor coils are placed very close to each other and a varying external voltage is applied to one of them (the primary coil), the induced magnetic field in this inductor reaches the second inductor (the secondary coil). The second inductor will then induce a varying voltage across itself. The outcome of using two inductors like this means that an applied voltage at the primary V_p will cause an induced voltage at the secondary V_s without there being a direct electrical connection between the two. To enable the magnetic linking between the coils to be improved, each coil is wound on to an iron core which is common to both. You may be wondering at this stage where the required ability to reduce the voltage comes from. This is the important part: the size of the induced secondary voltage V_s is dependent on the number of turns of wire N_s making up the secondary coil. More turns of wire mean a larger V_s. The size of V_s is also dependent on V_p and the number of turns of wire N_p making up the primary coil. It is in fact the ratio $N_s : N_p$ which describes the ratio $V_s : V_p$. These parameters are related in the following way

$$N_s/N_p = V_s/V_p$$

Note that this is true only if the voltage V_p is varying constantly as in a sine wave voltage

EXAMPLE 7.7

If the secondary coil of a transformer has 20 times more turns than the primary coil and $V_p = 1\ V_{rms}$, determine the size of the secondary voltage V_s.

SOLUTION

$$N_p = n$$

Therefore

$$N_s = 20n$$

$$N_s/N_p = 20n/n$$

$$= 20$$

Therefore

$$V_s/V_p = 20$$

This tells us that V_s is 20 times V_p Therefore

$$V_s = 20 \times V_p$$

$$= 20 \times 1$$

$$= 20V_{rms}$$

Because the ratio N_s/N_p is equal to the ratio V_s/V_p, we have three choices of transformer:

1. $N_s < N_p$ which results in $V_s < V_p$. This is called a 'step-down' transformer.
2. $N_s > N_p$ which results in $V_s > V_p$. This is called a 'step-up' transformer (as in Example 7.7).
3. $N_s = N_p$ which results in $V_s = V_p$. This is called an 'isolating' or '1 : 1' transformer.

EXAMPLE 7.8

If the primary coil of a transformer has 10 times more turns than the secondary coil and $V_p = 230\ V_{rms}$, determine the size of the secondary voltage V_s.

SOLUTION

$$N_p = 10n$$

Therefore

$$N_s = n$$

$$N_s/N_p = n/10n$$

$$= 1/10$$

Therefore
$$V_s/V_p = 1/10$$

This tells us that V_s is 10 times smaller than V_p. Therefore

$$V_s = V_p/10$$

$$= 230/10$$

$$= 23V_{rms}$$

This is a step-down transformer.

If the magnetic linkage between the primary and secondary coils is good (and it usually is), then the output power P_{out} and input power P_{in} of the transformer are equal Therefore, since $P = V \times I$, the primary current I_p and the secondary current I_s are simply related by

$$V_p \times I_p = V_s \times I_s$$

giving

$$V_s/V_p = I_p/I_s$$

Therefore

$$N_s/N_p = I_p/I_s$$

In other words, a step-up transformer would give $V_s > V_p$, but would reduce the secondary current accordingly. Remember, it is not possible to use a transformer to increase the voltage and the current together; to do so would mean an increase in power which the transformer is incapable of producing.

7.15.2 The half-wave rectifier

The purpose of a rectifier circuit is to provide a voltage or current in one direction only, and the diode is the component which can do it.

If we take the circuit in Figure 7.13 and look at the voltage across R instead of across the diode, we will see that we have a circuit which provides a pulsed positive output waveform from a standard a.c. input. The a.c. input voltage would normally

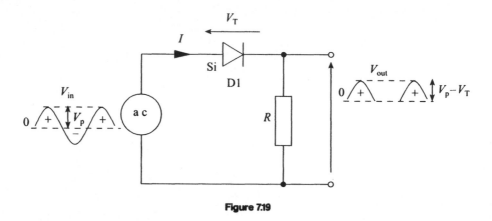

Figure 7.19

come from the secondary coil of a transformer. Figure 7.19 shows the circuit of Figure 7.13 redrawn, with the diode and the R positions interchanged. This is the half-wave rectifier, which is so called because it gives an output only when the input is on its positive half-cycle. We get no current flow during the negative half-cycle and therefore no output.

Note the following:

1. We lose V_T across the diode as normal, which leaves us with $V_{out} = (V_p - V_T)$.
2. During the negative half-cycle, the full V_p value is applied in reverse bias across the diode. If we are rectifying large voltages, we must be sure that we do not exceed the PIV rating for the diode.

Half-wave rectifiers do nothing during half the time, and therefore we could do with a circuit which is more efficient. Enter the full-wave rectifier circuit.

7.15.3 The full-wave rectifier (bridge rectifier)

The full-wave rectifier circuit uses four diodes arranged in a bridge circuit as shown in Figure 7.20 (the transformer secondary is omitted for clarity). The circuit operation is explained more easily if we split it into two halves which show the current flow through the circuits on each of the input half-waves. The current path for the positive half-wave is shown in Figure 7.21(a) and that for the negative half-wave is shown in Figure 7.21(b). There are four main points to note about this bridge circuit:

1. The current flows through two diodes on each half-cycle. Therefore, the voltage remaining across R_L is always $(V_p - 2V_T)$.
2. Whichever way the current flows, it is always forced through R_L in the same direction, producing a voltage pulse across R_L, also in the same direction.

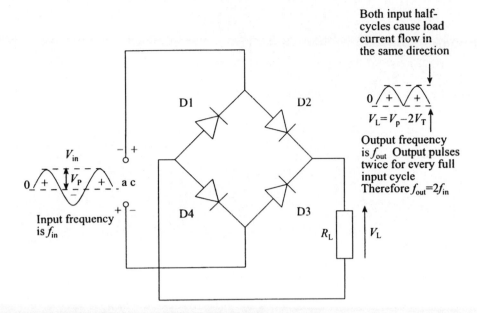

Both input half-cycles cause load current flow in the same direction

Output frequency is f_{out} Output pulses twice for every full input cycle Therefore $f_{out}=2f_{in}$

Frequency doubles

Figure 7.20

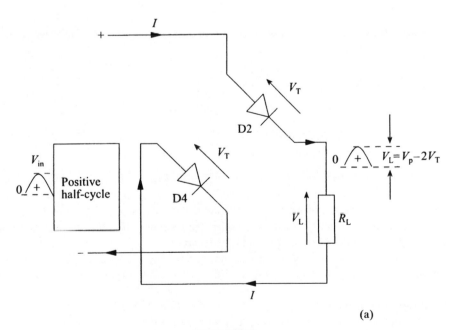

(a)

Figure 7.21(a)

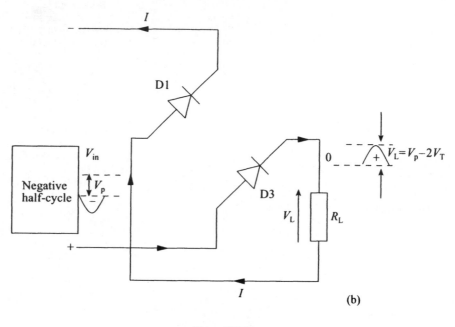

Figure 7.21(b)

3. Since we get a voltage pulse output for every input half-cycle, the output waveform repeats *twice* as often as the input waveform. In other words, the output frequency (f_{out}) is twice the input frequency (f_{in}). Therefore, we can say

$$f_{out} = 2f_{in}$$

4. The output is rectified but still not suitable for use as a d.c voltage source.

7.15.4 The full-wave rectifier using a centre-tapped transformer

The full-wave rectifier circuit as shown in Figure 7.22 uses only two diodes. The slight difficulty with this circuit is the special centre-tapped transformer secondary required. Centre-tapped merely means that an extra wire is connected to the middle of the secondary coil and is made available. Transformers like this are not rare but are found less commonly than the untapped versions.

The centre-tapping means that the secondary coil output voltage V_s is split into two, as shown in Figure 7.22, giving a voltage of $V_s/2$ between the centre-tap and each end of the coil. Note that the polarities of the two half coils (shown by arrows) are in the same direction.

During the positive half-cycle, point A will be more positive than the centre-tap and point B will be more negative than the centre-tap. This results in diode D1

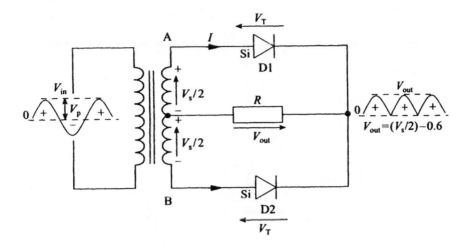

Figure 7.22

conducting and current flowing through R_L Diode D2 will be reverse biased, and therefore no current will flow through it.

During the negative half-cycle, point A will be more negative than the centre-tap and point B will be more positive than the centre-tap. This results in diode D2 conducting and current flowing through R_L. Diode D1 will be reverse biased, and therefore no current will flow through it. Note that the current flow through R_L is in the same direction as before.

The R_L voltage will have the same polarity during both input half-cycles to the circuit, its maximum value being equal to $V_s/2$ minus one diode drop, i.e

$$V_{out} = [(V_s/2) - V_T]$$

Because it is a full-wave rectifier, as with the bridge circuit, the output frequency is double the input frequency

7.15.5 The smoothing circuit

A smoothing circuit can be attached to the outputs of either the half-wave or the full-wave rectifier, but it is easier to smooth the full-wave output. We will consider smoothing the bridge rectifier output The simplest (but not the only) smoothing circuit consists of nothing more than a single capacitor in parallel with the load resistance R_L The circuit shown in Figure 7.23 works by charging C in bursts from each pulse of rectifier output, while the load resistance is continually discharging C. This repeating charge/discharge of the voltage across C produces a rippling voltage which, for a number of applications, is quite steady enough to use directly as a d.c. voltage! Typically, a ripple voltage (v_r) would be specified as a percentage of the

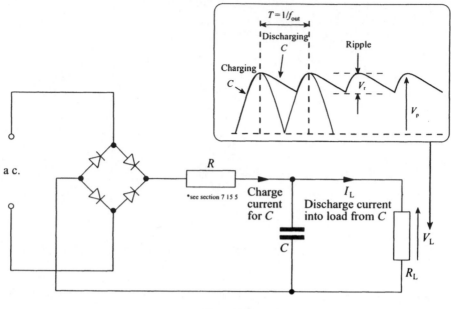

Figure 7.23

required output voltage V_L. For example, a 10% ripple on a 10 V output would normally mean that the V_L can vary from 10 V down to 9 V (but no lower).

The resistance R is added in some circuits to protect the diodes from the large surge current flow when the capacitor initially charges. When the first cycle is complete, the charge current is much smaller since it is only required to top up the capacitor voltage from $(V_p - v_r)$ to V_p. On power-up, the capacitor surge current will be limited only by the two (for a bridge circuit) diode resistances $(2R_D)$ and any transformer output coil resistance $(R_{transformer})$, giving

$$I_{surge} \approx V_L/(2R_D + R_{transformer})$$

These resistances will *never* total less than a few ohms, but it is worth checking the diode curve for an R_D value if the power supply is giving a large output voltage. We can safely assume for most cases that

$$(2R_D + R_{transformer}) \geq 2\,\Omega$$

For many applications this R can be left out, provided that the diode can handle the short-lived surge current. In diode data sheets you will find that a surge current rating is quoted which is many times larger than the recommended steady current value.

Finding the value of C
You can see that the circuit required to smooth the output waveform is simple How

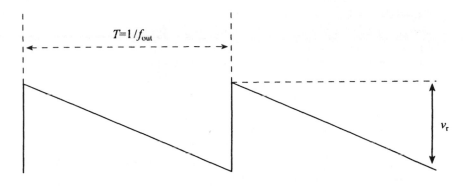

The ouput waveform simplified to allow easier analysis

Figure 7.24

about finding a value for C? This can easily be done if we make sensible use of some simplifications to the theory. First, we simplify the output waveform to make it easier to analyse. This is shown in Figure 7.24.

Look at Figure 7.24 and then follow these steps:

1. Assume that the discharge current (I_L) into the load resistance is linear. This current reduces the output from V_p to $(V_p - v_r)$: a voltage change of v_r.
2. Assume that the discharge current flows for a time (T) equal to the time interval between voltage peaks. (It is slightly less than this really.)
3. Use the expression relating capacitor current to rate of change of voltage across its plates, i.e. $I = C \, dV/dt$
4 Rewrite this expression in terms of our smoothing circuit values, i.e. $I = I_L$, $dV = v_r$ and $dt = T = 1/f_{out}$
5. Rearrange this equation to find the value of C for any required ripple voltage or load current.

Our simplified equation is

$$C = I_L/(f_{out} \times v_r)$$

Quite often the value of I_L might not be known directly, but this is not a problem since we would probably know the required output voltage (V_L) and the load resistance value. I_L then simply becomes V/R. To be more precise, $I_L = V_L/R_L$, and therefore

$$C = (V_L/R_L)/(f_{out} \times v_r)$$

Let us see an example to piece together these ideas.

EXAMPLE 7.9

Design a bridge rectifier and smoothing circuit, for use in the UK, to provide a voltage of 20 V for a load drawing 1 A. 10% ripple is allowable. Find the size of the supply required and the value of the smoothing capacitor. Use silicon diodes

SOLUTION

Look at the circuit in Figure 7.23.

1. First find the a.c. voltage required as input to the bridge. The V_L required is 20 V. Therefore, $V_p - 2V_T = 20$ and using $V_T = 0.6$ V we get

$$V_p = 20 + 1.2$$

$$= 21.2 \text{ V}$$

We normally express a.c. supply voltages in r m.s. value, and therefore we have

$$V_{in(rms)} = 21.2 \times 0.707$$

$$= 14.98 \text{ V}$$

$$\approx 15 \text{ V}$$

2. Next we find the ripple voltage value. We are told that we are allowed 10% ripple. This means that

$$v_r = 10\% \times V_L$$

$$= (10/100) \times 20$$

$$= 2 \text{ V}$$

3. We are given the I_L value directly as $I_L = 1$ A.
4. Finally, we need f_{out}. We are told that the circuit is to be used with mains UK supply, and therefore

$$f_{in} = 50 \text{ Hz}$$

Hence

$$f_{out} = 2f_{in}$$

$$= 100 \text{ Hz}$$

This is everything we need to calculate C. Therefore

$$C = I_L/(f_{out} \times v_r)$$
$$= 1 \text{ A}/(100 \text{ Hz} \times 2 \text{ V})$$
$$= 0.005 \text{ F}$$

To be more professional we would quote the C value in more appropriate units

$$C = 5000 \text{ } \mu\text{F}$$

One final point
Assuming $(2R_D + R_{transformer}) \geq 2\Omega$ (mentioned earlier in the text), we have

$$I_{surge} \approx V_L/(2R_D + R_{transformer})$$
$$\approx 20 \text{ V}/2 \Omega$$
$$\approx 10 \text{ A}$$

We therefore use diodes with a surge current rating of at least 10 A

A final word should be said about all the approximations used in this analysis. We can accept the simplifying techniques used because they produce a theoretical C value which is over-generous. That is, the C value calculated would actually produce a slightly smaller ripple than the one needed, which means that V_L is well within the specifications. In the real world of large-value capacitors, we can only buy +/−20% components and these come in only a rather limited range of values, so there is little point in being more exact in our analysis!

Self-assessment questions

SAQ7.17 In Example 7.9 what would happen to the size of the ripple if·
1. C were increased?
2   I_L increased?
3   The supply frequency increased?

SAQ7.18 Design a bridge rectifier and smoothing circuit, for use in the UK, to provide a voltage of 50 V for a load drawing 100 mA 5% ripple is allowable Find the size of the supply required and the value of the smoothing capacitor. Use silicon diodes

SAQ7.19 Design a bridge rectifier and smoothing circuit, for use in the USA (mains frequency is 60 Hz), to provide a voltage of 25 V for a 100 Ω

load. 10% ripple is allowable. Find the size of the supply required and the value of the smoothing capacitor. Use silicon diodes.

SAQ7.20 The audible hum from the loudspeaker of mains-driven hi-fi systems and TVs comes from the ripple voltage on the power supply output. If bridge rectifiers are always used, what frequency is the note you can hear:
1. in Britain?
2. in the USA?

Although for many applications the smoothing circuit works fine, it is not useful if the load resistance is very variable, i.e. if I_L varies. From our smoothing capacitor equation

$$C = I_L/(f_{out} \times v_r)$$

we can see that

$$v_r = I_L/(f_{out} \times C)$$

which tells us that v_r increases if I_L increases. This is not good as it also causes the average output voltage to reduce (see Figure 7.25). We now need a way of regulating the output voltage.

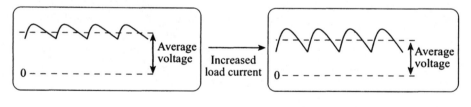

Figure 7.25

The next section deals with the last circuit in the power supply chain, which reduces the ripple to (almost) zero and removes the dependence of the size of the ripple on the load current. It is the zener diode voltage regulator.

7.16 Zener diodes

Armed with your knowledge of diodes and diode circuits, you will find this look at zener diodes will not present many problems Figure 7.26 shows its symbol and the component marking.

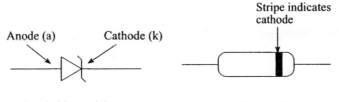

Figure 7.26

7.16.1 The zener voltage

The zener diode is a special diode which is manufactured to have a very low reverse bias breakdown voltage. A quick look at Figure 7.2 should remind you that the standard diode has a reverse breakdown value in the region of around 100 V. For the zener, however, this breakdown voltage is much reduced and the circuit designer can buy, off the shelf, a zener with a specific breakdown voltage from a large range of values. The breakdown voltage is called the 'zener voltage' (V_z).

7.16.2 Reverse bias

The most important fact about using zeners in circuits is that they must always be used *in reverse bias* This seems strange at first, but you will get used to it Let us look at a picture of the characteristic curve.

7.16.3 The zener diode characteristic curve

The zener diode characteristic curve is mostly identical to the standard diode curve except for the reduced (and very important) V_z value. The curve is shown in Figure 7.27 Note these three things about the zener.

1   It conducts in forward bias the same way as a standard diode, with a characteristic V_T voltage drop across it.
2   It also conducts in reverse bias when a voltage of V_z is applied
3   When it conducts in reverse bias, it does not conduct perfectly (what diode does?) but has a small resistance (R_z) which causes the curve to lean outwards slightly. This resistance is often called the 'incremental resistance' or the 'zener diode resistance' (R_z) Its value can be found in the same way that the average resistance (R_{av}) of a forward biased diode was found by finding $\Delta V/\Delta I$ on the curve.

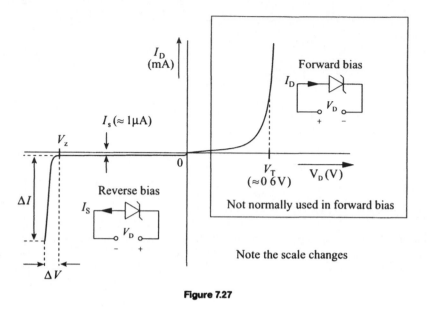

Figure 7.27

7.16.4 The real zener diode

Real zener diodes can be bought with a V_z chosen normally from one of the E12 resistance range of values, i.e. 10, 12, 15, 18, 22, 27, 33, 39, 47, 56, 68, 82. Multipliers are used to extend this range. For example, we could buy a zener with a V_z of 3.9 V, 39 V or 390 V The only other main parameter to watch out for is the zener power rating ($P_{z(max)}$). We can also buy zeners from a large range of different power ratings. When a zener is in reverse bias, it drops V_z and carries I_z. The power dissipation (P_z) is therefore $V_z \times I_z$ The zener dissipates power as heat and the maximum power allowed for a particular component must not be exceeded

$$P_{z(max)} = V_z \times I_{z(max)}$$

Since $P_{z(max)}$ is always quoted for a particular zener, we must make sure that $I_{z(max)}$ is not exceeded in our circuits.

7.16.5 Zeners as voltage references

Zeners really excel as voltage references. Whenever a voltage is required in a circuit which is less than the power supply value, we can use a zener This is a much better solution than using a voltage divider because the Thevenin resistance of a zener circuit can be made much lower and is therefore a much better voltage source (remember Thevenin equivalent circuits from Chapter 2). In Figure 7 28 we have a 6.8 V zener providing a steady 6 8 V output. Note that the zener is connected

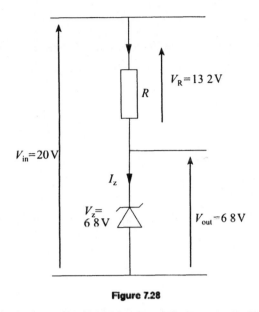

$V_{in} = 20\,V$

R

$V_R = 13\,2\,V$

I_z

$V_z = 6\,8\,V$

$V_{out} - 6\,8\,V$

Figure 7.28

in reverse bias and I_z flows through it from the cathode to the anode The beauty of this simple circuit is that if V_{in} is increased to 30 V, V_{out} stays fixed at 6 8 V. If you are not sure why, look at the characteristic. An increase in V_{in} merely causes R (the current limiter) to drop more voltage, which in turn causes the zener current I_z to increase. Provided that we do not exceed the power rating of the zeners, all is fine! Alternatively, if V_{in} is decreased, we still get $V_{out} = 6\,8\,V$

Imagine now that we require a 12 V output instead of 6.8 V The solution is simply to replace the 6.8 V zener with a 12 V one It could not be easier The zener voltage reference is a simple but tremendously useful little circuit.

7.16.6 Zener voltage regulators

At the end of Section 7.15 we had reached the point in the power supply where we needed to flatten out the ripple voltage from our smoothing circuit The solution to this is to use a voltage regulator, of which there are many types available. The really good ones are integrated circuits We will look first at the zener voltage regulator, as it also gives you an insight into using zeners generally The simple zener voltage regulator circuit is merely the circuit we have already seen in our look at zeners as voltage references. We will now place special emphasis on the current requirements of the load circuit. Look at Figure 7.29(a), where we are regulating a 20 V input to a 6.8 V output to drive a load resistance of 1 kΩ

From Figure 7.29(a) we can see that I_R always divides into I_L and I_z. Therefore, if I_L reduces, I_z increases. Similarly, an increase in I_L causes a reduction in I_z The relationship $I_R = I_L + I_z$ is always true

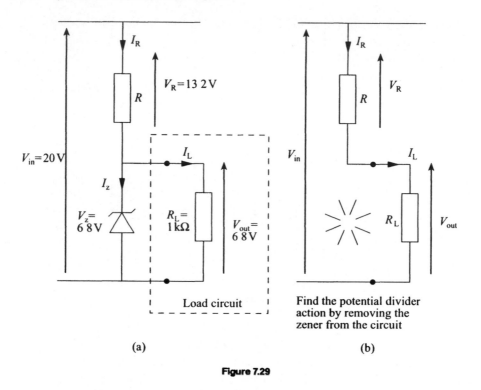

(a)   (b)

Figure 7.29

Now for the design process. Zener regulator design can be broken down into three steps, as follows:

1.   Make sure that the value of R_L is given or can be calculated.
2   For the zener to reach reverse breakdown it must receive an applied voltage equal to V_z. This means that the potential divider action of R_L and R across V_{in} must *not* reduce V_{out} to a value less than V_z. We check this by (mentally) removing the zener from the circuit, to leave the potential divider circuit. We always arrange for the voltage divider to give an output of at least V_z, i.e. $V_{out} \geq V_z$. Since

$$V_{out} = V_{in} \times R_L/(R + R_L)$$

we have

$$V_{in} \times R_L/(R + R_L) \geq V_z$$

This equation gives us the R value as everything else is known. When rearranged it gives us

$$R \leq [(V_{in}/V_z) - 1] \times R_L$$

3. Check the required power rating for the zener Remove the load from the circuit so that I_L is 0 This leaves us with an expression for I_R

$$I_R = I_L + I_z$$
$$= 0 + I_z$$
$$= I_z$$

Now that we know R we can find I_R from the voltage drop across R, where

$$I_R = (V_{in} - V_{out})/R$$

Therefore, with the load removed we have

$$I_z = (V_{in} - V_{out})/R$$

This is the maximum size of I_z for this circuit Therefore, we can write

$$P_z = I_z \times V_z$$

We would now choose a zener with a power rating of $P_{z(max)} > P_z$

Now let us design the zener voltage regulator for the circuit in Figure 7 29(a)

EXAMPLE 7.10
Find a suitable R value and zener power rating for the zener voltage regulator in Figure 7.29(a).

SOLUTION
1 R_L is given directly. If it is not, then find it from V_{out}/I_L

$$R_L = 1 \text{ k}\Omega$$

2. We are given $V_{out} = V_z = 6.8$ and $V_{in} = 20$ V. Therefore, from

$$R \le [(V_{in}/V_z) - 1]R_L$$

we have

$$R \le [(20/6.8) - 1] \times 1 \text{ k}\Omega$$
$$= 1.94 \text{ k}\Omega$$

Because real resistors have a +/− percentage error, we ensure that the zener will conduct with this resistance by choosing the next preferred resistor value below this, say

$$R = 1.8 \ k\Omega$$

3   If R_L is disconnected, we have

$$I_z = I_R$$

$$= (V_{in} - V_{out})/R$$

, Therefore

$$I_z = (20 - 6.8)/1.8 \ k\Omega$$

$$= 7.33 \ mA$$

Hence

$$P_z = V_z \times I_Z$$

$$= 6.8 \times 7.33 \ mA$$

$$= 49.6 \ mW$$

We would therefore use a power rating one up from this Hence

$$P_{z(max)} = 100 \ mW$$

which is fine!

Self-assessment questions

SAQ7.21 Design a zener diode regulator to give a 3.3 V output from a 10 V input for a 600 Ω load

SAQ7.22 Design a zener diode regulator to give a 12 V output from a 50 V input for a load drawing 500 mA.

SAQ7.23 Design a zener diode regulator to give a 3.9 V output from a 12 V input for a load drawing a varying current from 100 mA to 500 mA (Hint: use the smallest load resistance to find R.)

SAQ7.24 If a zener diode was running too hot, would you increase or decrease the load resistance?

SAQ7.25 What would V_{out} of Example 7 10 be if R was changed from 1.8 kΩ to 18 kΩ by mistake?

SAQ7.26 If the load was never disconnected from the zener voltage regulator circuit, could we reduce the power rating of the zener in Example 7.10? If yes, to what value?

Varying the input voltage of a zener voltage regulator
Notice that the zener voltage regulator in Example 7.10 has a fixed d.c input voltage You might wonder what happens if the input voltage is varying a little, as in our case of the power supply output voltage ripple The answer is that we use the same circuit, but we have to be more careful in our calculations of R and P_z
The extra design steps for a varying input voltage are as follows·

1 Find the maximum and minimum input voltage range ($V_{in(max)}$ and $V_{in(min)}$)
2. When finding the value of R, use $V_{in(min)}$. This will ensure that the smallest input will reverse bias the zener
3. When finding P_z, use $V_{in(max)}$. This ensures that we have enough zener power rating for the largest input voltage and hence the largest I_z value.

EXAMPLE 7.11
The circuit in Figure 7.29(a) is driven from the output of a 20 V smoothing circuit with a 10% ripple. Design the zener voltage regulator.

SOLUTION
1 $V_{in(max)} = 20$ V

$$V_{in(min)} = 20 - [(10/100) \times 20 \text{ V}]$$
$$= 18 \text{ V}$$

2 Using $V_{in} = 18$ V

$$R \leq [(V_{in}/V_z) - 1]R_L$$
$$\leq [(18/6.8) - 1] \times 1 \text{ k}\Omega$$
$$\leq 1.647 \text{ k}\Omega$$

Therefore, choose

$$R = 1.5 \text{ k}\Omega$$

3. Using $V_{in(max)} = 20$ V

$$I_z = (20 - 6.8)/1.5 \text{ k}\Omega$$

$$= 8.8 \text{ mA}$$

Hence

$$P_z = V_z \times I_z$$

$$= 6.8 \times 8.8 \text{ mA}$$

$$= 59.8 \text{ mW}$$

We would therefore use a power rating one up from this. Hence

$$P_{z(\text{max})} = 100 \text{ mW}$$

which is still fine!

Problems with zener voltage regulators

We cannot leave this section without commenting on four main problems of using zeners as voltage regulators:

1. The main problem is due to the incremental resistance of the zener. In other words, the diode does not conduct perfectly and (unlike our assumption in the examples given) does not therefore maintain a rigid voltage drop at different current flows. This means that, as the zener carries more reverse current, its voltage drop increases.
2   Zeners have wide tolerances (percentage inaccuracies) in V_z values, especially at low voltages. This means that we are never certain of the V_z value and replacing a damaged zener may result in a different voltage being produced.
3. The V_z values are temperature-dependent. Change the temperature and you change the voltage!
4. Integrated circuit regulators are a vast improvement on zener regulator circuits and are generally much easier to use (as we shall see).

Self-assessment questions

SAQ7.27 Design a zener voltage regulator driven by a smoothing circuit output of 10 V with a 10% ripple to give 4 7 V output to a load drawing 100 mA.

SAQ7.28 A certain CD player requires a steady 3.3 V supply at 150 mA It is to be driven from the cigarette lighter voltage supply in a car

Unfortunately, the car supply output voltage varies with the speed of the car: from 12 V to 15.5 V. Design a suitable zener voltage regulator circuit.

7.16.7 Integrated circuit regulators

As mentioned earlier, the really good voltage regulators are integrated circuit (i c) types. They usually have many advantages over the zener regulators, such as giving a more stable output, being easier to design with, having circuits which automatically limit output current, and having automatic thermal shut-down circuits (in case the i c gets too hot). Like all voltage regulators, the input voltage V_{in} must always be higher than the required regulated output voltage V_{out} (usually called the input–output differential voltage) by 2 or 3 volts. The V_{in} value usually has an acceptable range. For example, the LM7805 accepts V_{in} between 7.3 V and 35 V. All regulators have a maximum output current I_{max} which, if exceeded, causes the i c. to shut down temporarily. Most i.c. regulators are available in three-pin packages, making them easy to wire up. They are available in two types: fixed and variable. Let us look at these in turn.

Fixed voltage regulators

Fixed voltage regulators are manufactured with internal component values set to produce a single voltage For example, the LM78xx series use the last two digits to describe the regulated output voltage V_{out}, e.g LM7805 gives $V_{out} = 5$ V, LM7812 gives $V_{out} = 12$ V and LM7815 gives $V_{out} = 15$ V. The i.c. connections could not be simpler, having a single pin for V_{in}, V_{out} and common. The common pin is usually connected to earth. To demonstrate how easy they are to use, look at the circuit in Figure 7.30, where the LM7805 is used to produce 5 V at up to 1 A. Using the LM7805, for regulation to occur we must have $V_{in} \geq 7.3$ V.

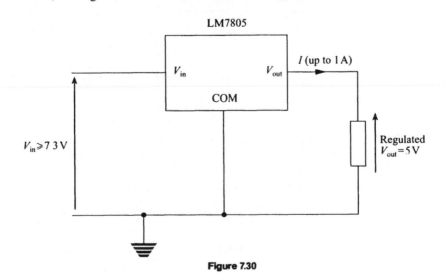

Figure 7.30

Variable voltage regulators

Sometimes the need arises to produce a regulated voltage which can be increased or reduced as and when necessary. The easiest way to produce different V_{out} values is to use a variable voltage regulator, such as LM317, LM723, LM338, LM350 or LM78HG. The three-pin i.c.s are still very simple to use, although extra components are required. For example, the LM317 can be used to give V_{out} values of 1.2 V–37 V, as shown in the circuit in Figure 7.31. The values of R_1 and R_2 can be changed to vary V_{out}.

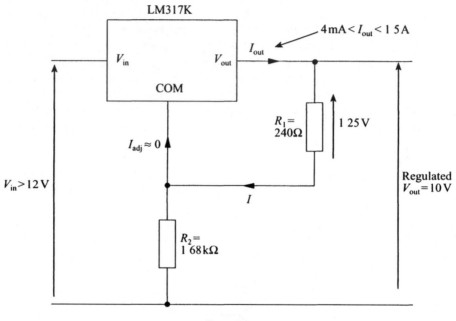

Figure 7.31

Using the LM317 as an example, let us see how the choice of resistor values is made. To do this properly, we need to know a little more about the i.c. The following four items of information are sufficient for use to design from·

1. The LM317 has three pins: V_{in}, V_{out} and ADJ. During its operation the regulator maintains a reference voltage of 1 25 V between V_{out} and ADJ (V_{out} being the positive end).
2. The input–output differential needs to be at least 2 V.
3. In operation the i.c. requires a minimum output current flow I_{out} of about 4 mA. The maximum I_{out} value is governed by the type letter on the i c. For example, the LM317K has $I_{out} = 1.5$ A.
4. The ADJ connection draws a negligibly small current I_{ADJ} of about 50 μA We assume, therefore, that I_{ADJ} is about 0.

Calculating V_{out}

With reference to Figure 7.31 we can say that since I_{ADJ} is about 0, the current in R_1 and the current in R_2 is equal to I. Therefore

$$V_{out} = I \times R_2 + 1.25$$

We also know that

$$I = 1.25/R_1$$

Substituting I into the equation for V_{out}, we have

$$V_{out} = [(1.25/R_1) \times R_2] + 1.25$$
$$= 1.25(R_2/R_1 + 1)$$

We can see from this that by choosing suitable R_1 and R_2 values we can produce a whole range of output voltages. There is, however, one limiting factor in this choice. R_1 must first be chosen to satisfy the condition of the minimum $I_{out} \approx$ 4 mA. Since we know that R_1 always has a 1.25 V across it, we can see that if $R =$ 240 Ω (the usual choice for this), $I = 1.25/240 = 5.2$ mA Since I is greater than 4 mA, we have satisfied the condition adequately.

EXAMPLE 7.12

Design a regulator circuit using an LM317 regulator to give an output of 10 V.

SOLUTION

First we choose $R_1 = 240\,\Omega$ to satisfy the minimum I_{out} condition. Next we rearrange the V_{out} equation to make R_2 the subject Therefore, we get

$$R_2 = [(V_{out}/1.25) - 1] \times R_1$$

Inserting the quantities for V_{out} and R_1, we have

$$R_2 = [(10/1.25) - 1] \times 240$$
$$= 1.68 \text{ k}\Omega$$

We must also ensure that $V_{in} > (V_{out} + \text{input–output differential})$, i.e.

$$V_{in} > 10 + 2$$
$$> 12 \text{ V}$$

If we required a continuously variable V_{out} (as in a laboratory power supply unit), we could easily make R_2 a variable resistor, remembering of course that we must ensure that V_{in} is about 2 V bigger than whatever V_{out} becomes!

Self-assessment questions

SAQ7.29 Design a regulator circuit using an LM317 regulator to give an output of 5 V. Specify the requirements of V_{in}.

SAQ7.30 Give three reasons why i.c. voltage regulators are better than zener voltage regulators.

SAQ7.31 When using an LM317 regulator, why would a choice of $R_1 = 1.25\,k\Omega$ be a bad idea?

SAQ7.32 Using an LM317 regulator, $R_1 = 240\,\Omega$ and $R_2 = 4.8\,k\Omega$. What size is V_{out} and what is the minimum V_{in}?

Checklist

Can you now:
- explain the difference between a standard and a zener diode?
- analyse simple diode circuits using static, dynamic and average resistances where needed?
- use LEDs in circuits?
- design simple voltage-clipping circuits?
- interpret existing voltage-clipping circuits?
- design rectifiers and smoothing circuits to fit a set of specifications?
- fully design a regulated power supply to fit a set of specifications?
- design voltage regulator circuits using fixed or variable i c chips?
- create reasonably accurate voltage references in a circuit?
- see why diodes are important and useful electronic devices?

Tutorial questions

1   A diode circuit includes other components with resistances of $2\,\Omega$ and $4\,\Omega$. Would it be sensible to consider the diode resistance value in any calculation?
2   Which diode resistance would we use if the circuit were driven by:
   1. a d.c. voltage?
   2. a large a.c. signal?
   3 a small a.c. signal?

3 A diode circuit is driven by a 50 V supply. Would it be sensible to consider V_T of the diode in any calculation?

4 Using Figure 7.6(a), what would be the diode current for a circuit with $V_{in} = 5$ V, $R = 80\,\Omega$ and $R_d = 10\,\Omega$?

5 Using Figure 7.6(a), what current limiting resistance would be used to limit the silicon diode current to 600 mA for a circuit with $V_{in} = 15$ V and $R_d = 5\,\Omega$?

6 If $I_s = 350$ nA for a diode at 50 °C, what value would the I_s of the diode reach if the temperature changed to:
   1 10 °C?
   2 80 °C?

7 If $V_T = 690$ mV for a diode at 23 °C, what value would the V_T of the diode reach if the temperature:
   1 decreased to 0 °C?
   2 increased to 60 °C?

8 What happens to the value of R_d when the diode current decreases?

9 What is the dynamic resistance value of a diode carrying a steady current, at room temperature, of·
   1 20 μA?
   2. 300 mA?
   3 150 nA?
   4. 1 A?

10 What does the PIV rating of a diode tell us?

11 What happens to the brightness of the LED if we halve the size of the current limiting resistor?

12 A 3 V battery is to be used to light a green LED whose I_{max} is 35 mA. Design the circuit.

13 The current used in calculating the answer for question 12 should not have been 35 mA. Why is this? (If you did use 35 mA, then it is time to pause for thought along the lines of being able to buy an exact value resistor.)

14 How can you protect an LED from a large reverse bias voltage?

15 Design an LED circuit to indicate when a 12 V battery has been connected correctly or the wrong way round to a circuit. Assume $I_{max} = 35$ mA.

16 Design a silicon diode voltage clipper to clip voltages more positive than +10 V and more negative than −0.6 V.

17 Design a silicon diode voltage clipper to clip voltages more positive than +1.2 V and more negative than 0 V.

18 If the secondary coil of a transformer has 50 times fewer turns than the primary coil and $V_p = 100$ V$_{rms}$, determine the r.m.s. and the peak voltage values of the secondary coil.

19 A certain 20 : 1 transformer is used with $V_p = 200$ V. The maximum rating of I_s is 2 A. Find the maximum value of I_p. Assume that all quantities are r.m.s.

20 If the input to a bridge rectifier is $25V_{rms}$ at 60 Hz, determine the peak output voltage and the output frequency

21   Design a bridge rectifier and smoothing circuit, for use in the UK, to provide a voltage of 6 V for a load drawing 0.75 A 12% ripple is allowable. Find the size of the supply required and the value of the smoothing capacitor. Use silicon diodes.

22   Design a bridge rectifier and smoothing circuit, for use in the USA (mains frequency is 60 Hz), to provide a voltage of 3 V for a 400 Ω load. 5% ripple is allowable. Find the size of the supply required and the value of the smoothing capacitor. Use silicon diodes

23   Why can we not earth both the lower end of the secondary output of the transformer *and* the lower end of the bridge rectifier output at the same time (as, for instance, when trying to view both these waveforms on a two-channel oscilloscope)?

24   Design a zener diode regulator to give a 9.1 V output from a 20 V input for a 1 5 kΩ load.

25   Design a zener diode regulator to give a 3 6 V output from a 30 V input for a load drawing 5 A.

26   Design a zener diode regulator to give a 5.1 V output from a 24 V input for a load drawing a varying current from 1 mA to 100 mA. (Hint: use the smallest load resistance to find R.)

27   Repeat question 26, but this time add the extra complication that the input voltage may vary from 24 V to 30 V.

28   Design a circuit using an LM317 regulator to give an output of:
   1.   6 V;
   2.   18 V.
   Specify the requirements of V_{in}.

29   With reference to Figure 7.31, if $R_1 = 100\,\Omega$, determine the current in R_2. Does the value of R_2 alter the current flow through it?

30   Using an LM317 regulator, $R_1 = 150\,\Omega$ and $R_2 = 1.2\,k\Omega$. What size is V_{out} and what is the minimum value of V_{in}?

Further reading

Bogart T F. (1986) *Electronic devices and circuits*, sections 3 2–4. Merrill, Columbus, Ohio

Floyd T L (1988) *Electronic devices*, 2nd edn, chaps 3 and 4 Merrill, Columbus, Ohio

Horowitz P & Hill W (1987) *The art of electronics*, sections 1 17, 1 25–31, 5 01–3, 5 13, 5 15–18 Cambridge University Press, Cambridge, England

Millman J & Halkias C (1988) *Integrated electronics*, chaps 2 and 3 McGraw-Hill, Singapore.

CHAPTER 8

Introduction to bipolar junction transistors

Aims and objectives

The aim of this chapter is to give you the confidence to design and use bipolar junction transistors in simple configurations Although many electronic designs will consist of integrated circuits (either digital or analog), you will occasionally need to find a design which requires the versatility of a simple bipolar junction transistor circuit to increase voltage or current levels, usually for interfacing purposes. Study in this area will also give more insight into the workings of commercially available integrated circuits. This is of particular use when the information for a specific integrated circuit is given in schematic form and shows the discrete component design used for the layout of the circuit.

After studying this chapter you will be able to:

- *describe the action of the bipolar junction transistor*
- *state and explain the three modes of operation of the bipolar junction transistor*
- *design a simple bipolar junction transistor switch circuit*
- *draw the three bipolar junction transistor circuit configurations and identify their main characteristics*
- *explain what biasing means and why it is used*
- *draw and explain four different biasing techniques*
- *design in detail the voltage divider biasing method*
- *design the biasing for common emitter and emitter following circuits*

8.1 Introduction to bipolar junction transistors

The bipolar junction transistor is usually called a 'BJT' and is one of a family of transistor devices Its name is derived from the conduction process inside the device, which is a combination of two types of charge carrier of different polarity, bi meaning two and polar referring to the polarities. The charge carriers in question are electrons and holes. As in the case of the chapter on diodes, it is not our purpose here to explain the solid state physics required to understand the BJT conduction mechanisms, but to understand the BJT action and how we can control its operation.

8.2 The symbol and the real thing

First of all, the BJT is a three-terminal device and we have a choice of two types. Figure 8.1 shows the symbols for each type, the semiconductor geometry and one of the common packaging forms it takes in real life. As with the diode, the real thing has a thousand disguises, different sizes, shapes and even colours!

The two types are called 'npn' and 'pnp' which describe the types of semiconductor used in the semiconductor sandwich. If you look at the semiconductor layout (which is an approximation of the geometry of the actual device), you will see that a p-type chunk of semiconductor is sandwiched between two n-type pieces. The symbol shows the direction of current flow in the emitter. Note the difference in current direction for the npn and the pnp. The three connectors have names: base (b), emitter (e) and collector (c). The emitter is so-called because it emits charge carriers and the collector collects charge carriers. Not all the emitted carriers are collected because some of them (about 1%) recombine in the base region (electron–hole pairs).

BJTs can be manufactured from a variety of semiconductor materials (just like diodes). The main contender is the silicon BJT, which is the one we will assume we are using throughout this chapter.

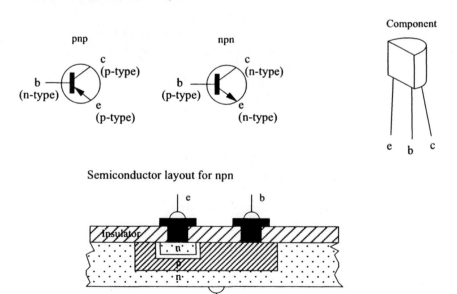

Figure 8.1

Although BJTs are either npn or pnp, their actions can be explained in identical ways To prevent a duplication of our investigation into BJT circuits, we will consider here only the npn BJT All the explanations and arguments for an npn can be used for a pnp, by swapping around the voltage polarities and reversing the current flows whenever these are shown

8.3 Back-to-back diode model

The BJT appears to be two p–n junction diodes connected together at the base. This model can be useful when testing a BJT before it is inserted into a circuit. It is a useful out-of-circuit test as each diode (or BJT junction) can be tested separately. In Figure 8.2 the diodes are shown and the tests are listed As you can see, we would not expect a test like this to give a low resistance between the collector and emitter because the current flow is blocked in both directions by the diode action. This back-to-back diode model is of use only in this context As we shall see shortly, when we connect the BJT into a circuit we get quite a different effect.

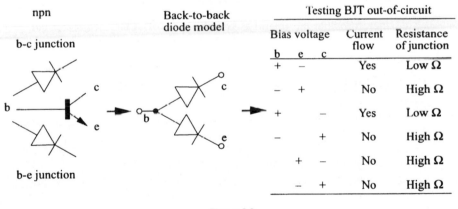

Bias voltage			Current flow	Resistance of junction
b	e	c		
+	−		Yes	Low Ω
−	+		No	High Ω
+		−	Yes	Low Ω
−		+	No	High Ω
	+	−	No	High Ω
	−	+	No	High Ω

Figure 8.2

8.4 BJT basics

We can now investigate how the BJT performs when it is in-circuit. First, since the b–e junction is a silicon diode we must forward bias it before we can obtain any current flow When we do this we also get current flow from the collector to the emitter. This is where our simple back-to-back diode model breaks down. Let us therefore describe the BJT action in a few simple rules with reference to Figure 8.3:

1. To obtain a current flow into the base, we must forward bias the b–e junction. To do this we apply a voltage at the base which is 0.6 V higher than that at the emitter. This voltage is called the base-emitter voltage (V_{be}).

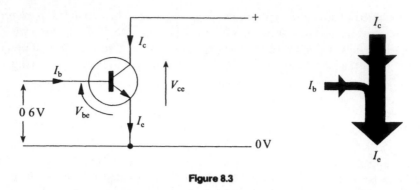

Figure 8.3

Therefore, when $V_{be} = 0.6\,\text{V}$, base current (I_b) flows.

2. If a voltage is applied between the collector and emitter (V_{ce}), then the base current flow (I_b) causes a much larger collector current (I_c) and emitter current (I_e) to flow.
3. I_c flows into the BJT and I_e flows out of the BJT.
4 If we increase I_b, we get an increase in I_c (and I_e).

We can see, therefore, that the small base current controls a large collector current For this reason the BJT is a current amplifier.

There are a few BJT expressions to remember:

1. The currents flow down through the BJT (as shown by the arrow on the BJT symbol)

$$I_e = I_c + I_b$$

Base and collector currents join up to form emitter current.
2. I_b is always much smaller than I_c ($I_b \ll I_c$). Therefore, we often use the approximation

$$I_c \approx I_e$$

3. For each BJT the ratio of I_c to I_b is an approximate constant called the current gain and is represented by β or h_{FE}. Therefore

$$\beta = h_{FE}$$

$$= I_c/I_b$$

The data sheets for any BJT will quote a range of values for h_{FE} because the manufacturing process cannot produce a BJT without slight variations in the structure. Typically, $h_{FE} \approx 100$, although different BJTs can have anything from $h_{FE} = 20$ (usually power transistors) to 1000.

EXAMPLE 8.1

What is the approximate size of I_c if a BJT with $h_{FE} \approx 100$ has

1 $I_b = 0$?
2 $I_b = 1\,\text{mA}$?
3. $I_b = 10\,\text{mA}$?

SOLUTION

Using $h_{FE} = \beta$

$\qquad = I_c / I_b$, we have

$\qquad I_c = I_b \times \beta$

Therefore

1. $I_c = 0 \times 100$
    $\quad = 0$

    i.e. there is no collector or emitter current flow.

2 $I_c = 1\,\text{mA} \times 100$
    $\quad = 100\,\text{mA}$

3. $I_c = 10\,\text{mA} \times 100$
    $\quad = 1000\,\text{mA or 1 A}$

8.5 V_{be} controls I_e

We can now look at BJT action in a little more detail. If I_b controls the size of I_c, you might ask how we control I_b The answer is that the amount of base current flow is dependent on the applied V_{be} value. In other words, once V_{be} reaches 0 6 V, a further tiny variation of V_{be} causes the collector current to vary quite a lot! The relationship between V_{be} and I_c (or I_e) is exponential, i.e. a small increase in V_{be} causes a large increase in I_c and I_e It is no surprise to find that the relationship between V_{be} and I_e is the same as the diode equation! The relationship is

$$I_e \approx I_s \times \exp(V_{be}/K)$$

The only difference this time is that we have V_{be} instead of V_D and I_e instead of I_D. K is still a constant of about 0.026 V at room temperature In fact, an increase in V_{be} of only 18 mV will cause the emitter current to double! It is because of this effect that the BJT can be used as a device to amplify voltage.

Let us have a word about the practicalities of this V_{be} value. We cannot keep increasing V_{be} because the BJT current flows would become too large and damage the device We therefore make sure that we limit the current flow to the BJT by introducing a current-limiting resistor. Typically, V_{be} will always lie within the range 0.5–0 8 V. Some power transistors have V_{be} reaching 1 V, but little more. (This is because of the extra internal voltage drops inside the BJT caused by the large currents involved.)

Self-assessment question

SAQ8.1 You have a radio receiver circuit which does not work. There are 6 npn BJTs in the circuit. You start to measure voltages across each BJT with your voltmeter The V_{be} of the first three BJTs is 0.5 V, 0 6 V and 0.7 V The V_{be} of the last three BJTs is 9 V, 0 7 V and 3 V. Which BJTs have blown?

8.6 BJT modes of operation

Now we know a little about the BJT, we can look at the way it behaves in a circuit. The device in action will *always* be operating in one of three distinct modes; these are cut-off, linear and saturation. Look at Figure 8.4, where a BJT is shown connected to a power supply voltage V_{cc} via a collector resistance R_c. This circuit is an example of one of the common BJT configurations used. It is called a 'common emitter' circuit because the input and output parts of the circuit are both connected by a common terminal to the emitter. It is easy to find out which mode a BJT is in it is all to do with the size of the collector-emitter voltage V_{ce}. Let us see how.

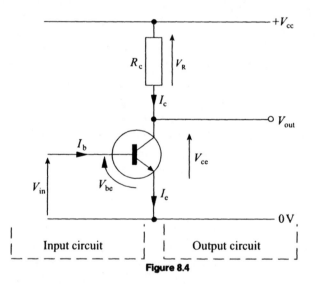

Figure 8.4

8.6.1 Cut-off mode

If $V_{in} < 0.6$ V, then the BJT is not forward biased and no I_b flows Therefore

$$I_c = I_e = 0$$

This means that there is no voltage drop across R_c and that $V_{ce} = V_{cc}$ This condition is known as 'off' or 'cut-off'. Because no I_c or I_e current flows, the collector–emitter terminals appear to act like an open switch. In fact, cut-off is one of the two modes used when the BJT is connected to act as a switch (see Figure 8 5).

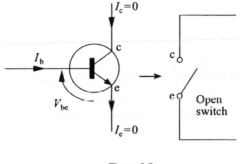

Figure 8.5

8.6.2 Linear mode

We now consider applying $V_{in} \approx 0.6 \, \text{V}$. This is sufficient to forward bias the b–e junction and cause I_b flow. The I_b flow in turn causes the much larger I_c and I_e to flow also You can see in Figure 8.4 that this will cause a voltage drop (V_R) to appear across R_c, reducing V_{ce}. Provided $V_{cc} > V_{ce} > 0.2 \, \text{V}$, we have the BJT working in its linear or active mode. We can say

$$V_{ce} = V_{cc} - (I_c \times R_c)$$

8.6.3 Saturation mode

The saturation mode describes the BJT when $V_{in} \approx 0.6 \, \text{V}$, the BJT is forward biased and I_b, I_c and I_e flow. This is similar to the linear mode, but for the size of V_{ce}. The I_c flow through R_c causes a voltage drop which reduces V_{ce} to about 0.2 V. We therefore have saturation if

$$V_{ce} \leq 0.2 \, \text{V}$$

There are some extra points about saturation to note here:

1. In saturation V_{ce} becomes $V_{ce(sat)}$ and I_c becomes $I_{c(sat)}$.
2. When $V_{ce(sat)}$ drops to about 0.2 V, the base-collector junction *also* becomes forward biased (Remember the diode behaviour?)
3. $V_{ce(sat)}$ for any BJT cannot drop much below 0.2 V, therefore any further increase in V_{be} will increase I_b, but I_c and I_e cannot increase as they are

limited by the minimum $V_{ce(sat)}$ value. This means that in saturation we may have the situation where $h_{FE} \neq I_c/I_b$. The extra I_b is wasted and serves only to increase the temperature of the BJT.

4.   Because of the very low $V_{ce(sat)}$ value, the collector-emitter terminals appear to behave like a closed switch (see Figure 8 6)

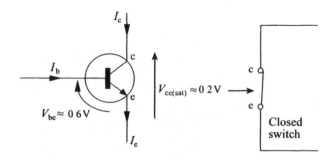

Figure 8.6

8.7 The d.c. load line

Summarizing these three modes of operation, we could say that a BJT is either in cut-off, in saturation or is somewhere in between, i.e. in the linear mode. To determine the mode we must have some information about the V_{ce} value. Referring to the common emitter circuit in Figure 8 4, we can easily plot a graph to show the relationship between the sizes of I_c and V_{ce} to illustrate the modes in a different way. This graph is called a 'd c load line'. From the circuit we can see that

$$V_{cc} = V_{ce} + (I_c \times R_c)$$

This is a linear expression and, if put in the form $y = mx + c$, we get

$$I_c = -[(1/R_c) \times V_{ce}] + (V_{cc}/R_c)$$

This equation tells us that if we plot I_c against V_{ce}, we will get a straight line with a slope equal to $-1/R_c$ and an intercept of V_{cc}/R_c. This graph is shown in Figure 8.7.
   Note that

1.   A different R_c would give a different slope.
2.   The region near the y intercept ($V_{ce} \approx 0\,V$) indicates the saturation mode.
3.   The arrow at the x intercept ($V_{ce} = V_{cc}$) indicates the cut-off mode.
4.   Any point on the load line between the intercepts indicates the linear mode.

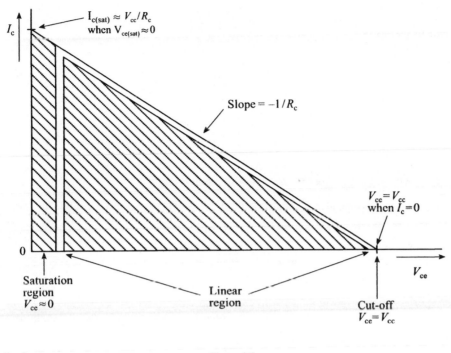

$I_{c(sat)} \approx V_{cc}/R_c$
when $V_{ce(sat)} \approx 0$

I_c

Slope $= -1/R_c$

$V_{ce}=V_{cc}$
when $I_c=0$

0

V_{ce}

Saturation
region
$V_{ce} \approx 0$

Linear
region

Cut-off
$V_{ce}=V_{cc}$

Figure 8.7

8.8 The BJT switch circuit

The BJT switch is one of the easier designs you will come across and is a good
starting point for demonstrating base current limiting and handling h_{FE} variations.

If you decide to switch the BJT directly from saturation to cut-off, the device
behaves very much like a switch with the switch contacts across the collector-
emitter. The switch, of course, is programmable from the V_{be} input: it is on when
$V_{be} \approx 0.6$ V and off when $V_{be} = 0$ V. This is the starting point of digital electronics!

Before launching into a switch design example, we will first look at the power
dissipation of BJT switches and the problems of using h_{FE} or β in your designs.

8.8.1 Power dissipation

In many applications the BJT is made to switch from on to off very quickly,
missing out the linear mode. When this happens the BJT has either a minimum
value of V_{ce} across it ($V_{ce(sat)} \approx 0$), or a minimum value of I_c ($I_c = 0$) through it.
Since power dissipation of the device is measured as

$$P = I_c \times V_{ce}$$

we either have

$$P = I_c \times 0$$

$$= 0$$

or

$$P = 0 \times V_{ce}$$

$$= 0$$

Hence these BJT switching circuits operate with minimum power dissipation

8.8.2 Current gain problems and worst case design

Current gain (β) values for a specific BJT type can be very unpredictable, for three reasons.

1   Mass production causes a 3 : 1 spread in the β value.
2. Temperature variations cause another 3 : 1 spread in the β value.
3   Increasing I_c decreases the β value.

These effects can result in a 10 : 1 spread in the value. Therefore, designs based on a specific β value are always avoided. Designers choose instead to always take the minimum quoted value for β. This is called 'worst case design'.

8.8.3 Designing the circuit

The design of the BJT switching circuit utilizes the saturation and cut-off modes we have discussed. We should be aware, however, that a 'switch' circuit does not necessarily have to switch something on: it can merely switch its output voltage from a high level to a low level. What the designer does with the switching voltage levels is a topic for a later study.

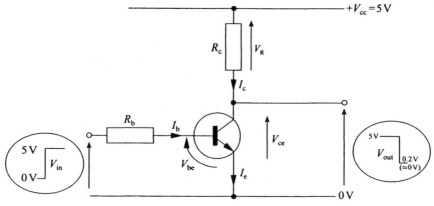

Figure 8.8

The switch circuit shown in Figure 8.8 switches voltage levels and is called a 'BJT switch', a 'logic inverter' or a 'logic translator', depending on its use The circuit action is as follows:

1. When V_{in} is 0 V, the BJT is in cut-off mode because there is insufficient V_{be} to forward bias the transistor No currents flow and $V_{ce} = V_{cc} = 5$ V.
2. When $V_{in} = 5$ V, the BJT starts to conduct. The V_{be} value cannot be more than ≈ 0.6 V and therefore the extra voltage is dropped across R_b, which also serves as a current limiter, where $I_b = (V_{in} - V_{be})/R_b$ The collector current flow causes a voltage drop across R_c. The resistor values are purposely chosen to ensure that this voltage drop ($I_c \times R_c$) is at least equal to V_{cc} Since V_{ce} cannot be less than ≈ 0.2 V, the current is limited to (V_{cc} $- 0.2)/R_c$ and $V_{out} = V_{ce(sat)} \approx 0\,2$ V.

Note what has happened here. A high input voltage causes BJT saturation and the output goes to almost zero volts. A low input voltage causes cut-off and the output goes to V_{cc}, i.e. V_{in} high causes V_{out} to be low and V_{in} low causes V_{out} to be high. We clearly have an inverting action here. A square wave would be inverted at the output You may wonder why it is called a logic translator The answer comes from the realization that V_{cc} does not have to be the same as the V_{in} high value. If the above circuit had $V_{cc} = 20$ V, then V_{out} would switch from 0.2 V to 20 V! (We would, of course, change the values of R_c and R_b to do this.) Therefore, an input square wave, which is the waveform generally used in digital logic circuits, can be inverted *and* amplified to a different amplitude This process is called 'logic translation' in some circles. Now let us look at a real switch design problem.

EXAMPLE 8.2

Design a BJT switch circuit to translate a 0–5 V square wave input to a 0–12 V square wave output An inverted output is acceptable The BJT specification is $I_{c(max)} = 100$ mA and $100 \leq \beta \leq 1000$.

SOLUTION

We will refer all our design steps to the switch circuit in Figure 8.8. First, we list the things we must take into account

1. The BJT must switch from cut-off to saturation, on 0 V and 5 V input levels respectively.
2. When $V_{in} = 0$, the BJT is off. We need $V_{out} = 12$ V Therefore, since $V_{ce} = V_{cc}$, we will make $V_{cc} = 12$ V.
3. When $V_{in} = 5$ V, BJT is in saturation We need $V_{out} \approx 0$ V. Therefore, choose the saturation voltage where $V_{ce(sat)} \approx 0.2$ V.
4. In saturation the $I_{c(sat)}$ must be less than the maximum current ($I_{c(max)}$) that the BJT can handle without damage. Therefore, we choose $I_{c(sat)} = 1$ mA (well away from maximum value!)

To find R_c, $I_{c(sat)} = 1$ mA and, in saturation

$$V_{Rc} = (V_{cc} - 0.2)$$
$$= 11.8 \text{ V}$$

Therefore
$$R_c = 11.8 \text{ V}/1 \text{ mA}$$

giving
$$R_c = 11.8 \text{ k}\Omega$$

(or the nearest preferred value).

To find R_b, we work backwards from the output to the input We know that when $V_{in} = 5$ V we have $I_c = 1$ mA. Therefore, $I_b = I_c/\beta$. We always use the worst case design and therefore the lowest β value. Using $\beta = 100$, we can calculate I_b

$$I_b = 1 \text{ mA}/100$$
$$= 10 \,\mu\text{A}$$

This current flows through R_b Therefore

$$R_b = (V_{in} - V_{be})/I_b$$
$$= (5 - 0.6)/10 \,\mu\text{A}$$
$$= 440 \text{ k}\Omega$$

That is all there is to it

Because we are using approximations, there is plenty of room for movement in these values In other words, we can choose R values near to the theoretical ones, but we need to consider the effect of this choice. If R_b is reduced in value, I_b increases but I_c is limited at ≈ 1 mA, and therefore the extra I_b is wasted. This means that we can choose any R_b value below 440 kΩ! We cannot, however, use a value bigger than 440 kΩ. Why not? Well, if R_b increases, I_b reduces and the BJT comes out of saturation. If this happens, V_{out} will no longer be at its minimum value. (There is a possibility that it will not be affected, but that is only if the β value was higher than 100, and we cannot be certain of that) The R_c value need not be exact, but a different value will require a new calculation of $I_{c(sat)}$, I_b and R_b.

We can also use this type of BJT switch to turn load circuits on and off This application of the switch is one of the most useful and can be rather interesting Take the case of a touch-sensitive light switch shown in Figure 8.9. We might have a 0.1 A 24 V lamp which is switched on and off via a touch plate consisting of two metallic contacts 1 mm apart. Using a BJT with $I_{c(max)} = 1$ A and $500 \leq \beta \leq 800$, the design calculations yield $R_c = R_{lamp} = 240 \,\Omega$ and $R_b \approx 117$ kΩ. We know that

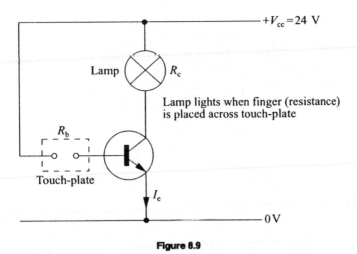

Figure 8.9

R_b can be 117 kΩ or less. A few experiments with fingers (at last a human element to all this!) will tell us that an average finger has a resistance across 1 mm of between 40 kΩ and 1 MΩ, depending on temperature, finger pressure and mental stress level (as perceived by a lie detector). This circuit will therefore light the lamp for most stressed-up liars who press hard on the plate! We can, of course, use a BJT with a higher β value for those who are honest, calm individuals, or we can lick our fingers before touching the plate. Note that the circuit takes its input voltage from the V_{cc} supply, this is quite common for switch circuits

Self-assessment questions

SAQ8.2 In Example 8 2, what would happen if the BJT actually had a β value of 1000? (Hint. what size of I_b would be required? Would we just be wasting more base current?)

SAQ8.3 Redesign Example 8 2 to give a 20 V output signal.

SAQ8.4 The base resistor of a switch circuit has gone open-circuit, causing failure of the circuit Its original value was 100 kΩ, but we only have replacement 47 kΩ and 1 MΩ resistors. Which would we use?

SAQ8.5 A 0 5 A motor is to be switched with a BJT What would be suitable $I_{c(sat)}$ and $I_{c(max)}$ values for the circuit?

SAQ8.6 If the R_c resistor in Example 8 2 were accidentally replaced by a lower value, what would be likely to happen to the power dissipation of the BJT and why?

SAQ8.7 Work through the calculation for the touch-sensitive switch circuit shown in Figure 8 9 to verify its operation

SAQ8.8 Design a circuit of your own which will allow us to turn a 12 V, 22 W car headlamp on and off from a cheap low-current switch mounted on the car dashboard Assume you have a BJT with $\beta = 800\text{--}1200$

SAQ8.9 The touch-sensitive switch circuit could easily be modified into a rain detector (either there would be water across the contacts or not) Design a rain detector to light a 40 mA LED when it rains on to the contacts (Remember the V_T value Assume rain-water resistance is about 40 kΩ)

8.9 Introduction to BJT linear circuits

The switch circuit has been discussed to illustrate the cut-off and saturation modes of BJT operation. There is a whole family of BJT circuits which rarely use either of these modes, but make extensive use of the linear mode. Before looking at detailed design steps for these circuits, we will look at their main circuit configurations and their main features. The circuits are broadly grouped into one of the following configurations. common collector (CC), common emitter (CE) and common base (CB). This is also the order of their popularity in common use. The circuit

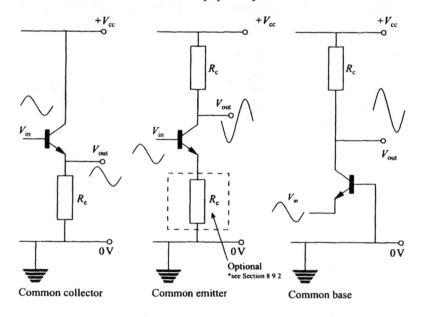

Figure 8 10

Table 8.1

Feature	Common collector (emitter follower)	Common emitter (emitter degenerated) (see Section 8 9 2)	Common base
A_v	≈ 1	High ($\approx R_c/R_e$)	High
A_i	High ($\approx \beta$)	High ($\approx \beta$)	≈ 1
A_p	High	Very high	High
R_{in} (Ω)	High ($\approx \beta \times R_e$)	Medium-high	Low
R_{out} (Ω)	Low	Medium-high ($\approx R_c$)	Medium-high ($\approx R_c$)
PC (°)	0	180	0

configurations are shown in Figure 8 10, where their differences can clearly be seen. Notice particularly where V_{out} is connected

Table 8 1 illustrates the differences in the three circuits. Let us have a quick look at the terms first. We have seen A_v before, but what about the others? A_i is the current gain of the circuit defined as $A_i = I_{out}/I_{in}$, where I_{out} is output current and I_{in} is input current; A_p is the power gain of the circuit defined as $A_p = P_{out}/P_{in}$, where P_{out} is output power and P_{in} is input power; R_{in} is input resistance, as normal; R_{out} is output resistance, as normal; and PC is phase change, defined as the amount of shift in phase between the input and the output.

8.9.1 Common collector configuration

The most common configuration is the common collector (CC). Usually called the 'emitter follower circuit', it has a voltage gain of about unity and provides no phase change. This means that the output follows the input in size and direction. V_{in} is applied to the base and V_{out} is taken from the emitter. There is usually no need for a collector resistance. Its name springs from the fact that the collector voltage does not change and is held at a common V_{cc} voltage. Looking at Table 8 1, you will notice that it has a high input resistance and low output resistance, making it an ideal contender for a buffer circuit For more information about buffers and impedance matching, take a look at the op-amp voltage follower circuit description in Chapter 3. It is used as a good voltage source within a circuit and as a power amplifier stage, since it has current gain. As we will see later, its input resistance can easily be approximated to $R_{in} \approx \beta \times R_e$, and therefore the designer can adjust the R_{in} to whatever is required by increasing or decreasing the value of the emitter resistance. The emitter follower circuit is very stable against temperature changes and is very predictable The reason for this stability is because R_e introduces negative feedback into the circuit (which is always a good thing for stability). This R_e feedback action will be covered in the BJT analysis sections.

We now consider how the circuit works. Look at Figure 8 11(a) and consider the following circuit action. When a voltage V_{in} of, say, 6 V is applied to the base, the

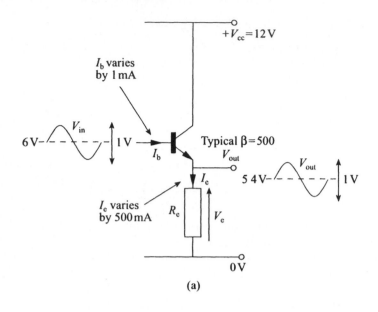

Figure 8.11 (a)

BJT receives sufficient forward bias to start I_b flowing. Remembering that V_{be} can never be more than 0.6 V, the emitter voltage (V_e) rises to ensure that this is true. V_e is now at 5.4 V, which determines the current flow (I_e) through the emitter resistor: $I_e = V_e/R_e$. This in turn determines the amount of I_b (which is always much smaller than I_e, since $I_e \approx I_c = \beta \times I_b$) required from V_{in}. The rest of the supply voltage (V_{cc}) is dropped across the BJT as V_{ce}. Now here comes the useful part. If V_{in} is now made to rise and fall sinusoidally around the average 6 V level, V_e will do exactly the same to ensure that V_{be} remains constant. The V_e variation also means that I_e will vary, which causes I_b to vary. The result is that a variation in V_{in} and I_b causes a much larger I_e variation through R_e The V_{in} and V_e variations are the same size and in phase. Therefore, we have a current gain of β and voltage gain of unity.

8.9.2 Common emitter configuration

The traditional (but rarely used) configuration has a fixed emitter voltage held at a common 0 V: hence its name. It is far more common to see the circuit in its more useful form with an extra emitter resistance added between the emitter and 0 V. This optional extra resistance turns the circuit into what is known as the 'emitter degenerated common emitter' (CE) circuit. You can see this in the CE configuration of Figure 8.10. All this name means is that in action it loses a small voltage across this emitter resistor when current flows, which reduces the amount of voltage available to be dropped across the V_{ce} of the BJT. The addition of this R_e introduces stability into the circuit via negative feedback action (like the

CC circuit) and is therefore preferable to a CE without the R_e The CE circuit is always used to boost an input voltage and therefore its voltage gain is important. Because of the negative feedback action of R_e, the A_v value is quite predictable and is approximated to the ratio of the collector and emitter resistance values

$$A_v \approx -R_c/R_e$$

(remember op-amp resistor ratios?). See the BJT analysis section for more about this The circuit amplifies and inverts, but its drawback is the large output resistance R_c. Therefore, to get more voltage gain, we also get a higher R_{out}

Let us see how the circuit works. Look at the CE configuration in Figure 8 11(b) The circuit action is identical to the CC circuit with the following extras·

1. The output is taken from the collector (not the emitter).
2 V_{in} variations cause I_e *and* I_c variations Unlike the CC circuit, the I_c variations are turned into voltage variations at the collector by the addition of R_c. R_c is really a current to voltage converter.
3 An increase in V_{in} causes an increase in I_c, which increases the voltage drop across R_c Since the top end of R_c is fixed at V_{cc}, the result is a drop in collector (or output) voltage The circuit therefore inverts
4 Increasing R_c increases the voltage drop across it, and therefore the size of the output voltage swing. The voltage gain expression also illustrates this

$$A_v \approx -R_c/R_e$$

The minus sign indicates the inverting action

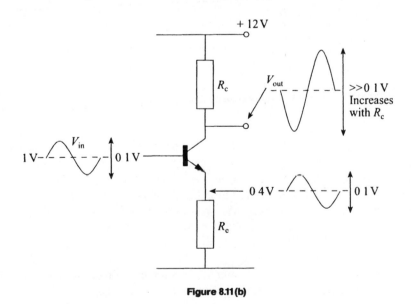

Figure 8.11 (b)

8.9.3 Common base configuration

With a low R_{in} and a high R_{out}, the obvious use of the common base (CB) configuration is as a load for a current source (remember the condition for good current transfer?) or as a load for a very low resistance voltage source. It is often seen as part of an oscillator circuit where the input to the emitter is from a low resistance transformer coil. The CB is also used in high frequency circuits because the capacitance at the emitter is considerably less than that at the base of a CE circuit. (A large capacitance across the input of a signal will have a small reactance as the frequency increases and therefore short-circuits the signal to the ground.) The CB circuit, having unity current gain and high voltage gain, is therefore quite useful. Its name, of course, describes the fixed base voltage to the 0 V line.

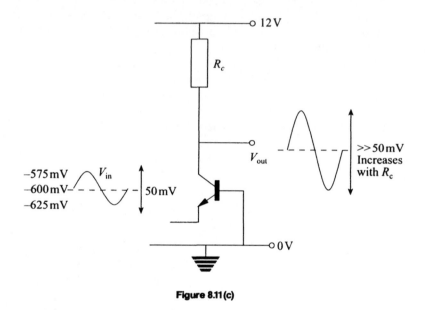

Figure 8.11 (c)

With reference to Figure 8 11(c), the circuit action can be briefly explained as follows. V_{in} needs to be below zero volts for the BJT to receive a forward bias, and therefore Figure 8.11(c) shows V_{in} held at a negative d.c offset of $-600\,mV$. Tiny variations of V_{in}, by a few millivolts, cause V_{be} to change by the same amount. As we have already discussed, varying V_{be} will vary I_e and I_c considerably. The I_c variation is converted to collector voltage variation by the presence of R_c. Larger R_c values give larger output voltages (as in the CE case). Increasing V_{in} causes a decrease in V_{be} and I_c, therefore reducing the voltage drop across R_c and increasing the collector (and output) voltage. V_{in} and V_{out} are therefore in phase Since $I_c \approx I_e$, and I_e is also the input current, there is a current gain of unity

Self-assessment questions

SAQ8.10 A BJT circuit has been set up with $I_e - 1$ mA and $I_b = 10\,\mu A$. What are the sizes of I_c and β?

SAQ8.11 Look at Figure 8 12 In this we added a collector resistor A voltage drop across the resistor develops when a collector current flows. Find the minimum value of R_c which would ensure that the BJT is in saturation. Notice that the current gain value is guaranteed to be only somewhere in the given range!

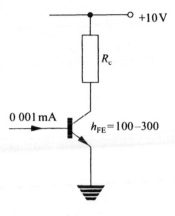

Figure 8.12

SAQ8.12 Roughly what size voltage would we expect to read at the base of the BJT in Figure 8.12?

SAQ8.13 Draw the d c load line of Figure 8 12 using the R_c value from the answer to SAQ8 11

SAQ8.14 If we required a circuit with unity voltage gain but high current gain, which configuration would we use?

SAQ8.15 What is another name for an emitter follower?

SAQ8.16 Which configuration gives us an 180° phase change?

SAQ8.17 Which configuration gives us a high voltage gain *and* a high current gain?

SAQ8.18 If we required an output voltage to perform as a good voltage source, which configuration would we use?

SAQ8.19 Why is the CB circuit no good for amplifying voltage signals from high resistance sources?

8.10 D.c. biasing

We will discuss the need for biasing and explain what it means by illustrating what happens if we do not use biasing.

8.10.1 Without a d.c. bias?

There are many occasions when we need to set up the BJT to work in its active or linear mode This is the most useful mode for amplifier design where analog signals need to be dealt with and we usually use one of the CE, CC or CB circuit configurations mentioned We have already discussed how the CC circuit configuration works, so let us see where biasing comes in or, more specifically at first, what happens when it is not there.

EXAMPLE 8.3

Design a CC circuit to accept a V_s of 1 V peak from a sine wave generator with a source resistance R_s of 1 MΩ to drive a 10 kΩ load R_L.

SOLUTION
First, an understanding of the problem is required (refer to Figure 8.13). The CC is needed because if the generator were connected across the 10 kΩ load directly, most of the voltage would be dropped across its internal source resistance of 1 MΩ. This is shown in Figure 8 13. Let us verify this. Voltage across the load V_L is

$$V_L = V_s \times R_L/(R_L + R_s)$$

$$= 1 \text{ V} \times 10 \text{ k}\Omega/(10 \text{ k}\Omega + 1 \text{ M}\Omega)$$

$$= 9.9 \text{ mV}$$

This indicates poor voltage transfer.

Therefore, we buffer the circuit with a CC and use R_L as the emitter resistance Assuming our BJT has $\beta = 500$, the CC input resistance $R_{in} \approx \beta \times R_L$. Therefore

$$R_{in} \approx 500 \times 10 \text{ k}\Omega$$

$$\approx 5 \text{ M}\Omega$$

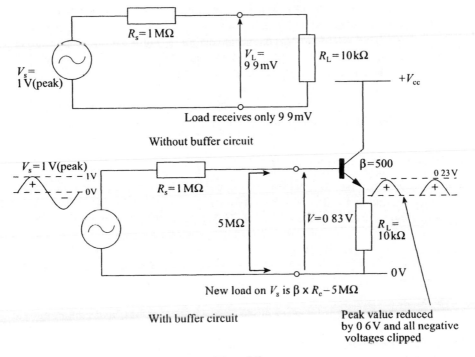

Load receives only 9 9mV

Without buffer circuit

New load on V_s is $\beta \times R_e - 5\,M\Omega$

With buffer circuit

Peak value reduced by 0 6 V and all negative voltages clipped

Figure 8.13

The load on the generator circuit is now $5\,M\Omega$. Therefore, when the CC is connected to the generator, the base voltage V_b will vary by

$$V_b = V_s \times R_{in}/(R_{in} + R_s)$$

$$= 1\ V \times 5\ M\Omega/(5\ M\Omega + 1\ M\Omega)$$

$$= 0.83\ V$$

Since we know that the CC has a voltage gain of unity, this 0.83 V peak sine wave signal should appear across the $10\,k\Omega$ load. (This is very similar to the op-amp voltage follower example in Chapter 3.)

So we appear to have solved the problem, but this is far from true Look at V_{out} in Figure 8 13. You can see that it is nothing like a 1 V peak sine wave! We still have a problem. Why? The answer is simple and twofold

1   The BJT will only go into forward bias when V_{in} reaches 0.6 V. Therefore, whenever V_{in} is below this voltage, the BJT is in cut-off mode, no currents flow and R_L has zero voltage across it for half the time

2. The second problem is that V_{in} must provide the required 0.6 V forward bias voltage, which results in V_e always being 0.6 V below V_{in}. The maximum V_e voltage, therefore, is not 0.83 V (the peak input voltage), but $(0.83 - 0.6) = 0.23$ V (see Figure 8 13). This circuit needs some modification before it can be used in this way.

The modification required to the circuit in this example is to add some d.c. bias to the BJT. Read on to find out how this is done.

8.10.2 Adding a d.c. bias

We can see clearly the effect of not having a d.c. bias from the previous example, but what does adding a d c. bias actually mean?

Generally, biasing something means to influence something continuously in one direction only. In electronics it means to add something continuously to force a device or circuit to be constantly turned on. In real terms it means to add a voltage or a current to enable a device to operate continuously without the need to add an extra external signal. When a BJT is biased, it means that it is in its linear mode, which is somewhere above cut-off but below saturation.

The way this is done is by ensuring that some collector, base and emitter current is flowing through the BJT at all times, even when there is no externally applied input signal. More often than not we choose component values so that the BJT is exactly half way between being cut-off and saturation. This means that the voltage dropped across the collector-emitter of the BJT (V_{ce}) is about half of the supply voltage. This is called 'mid-point' or 'mid-rail' biasing. So if the circuit supply is 12 V, we make sure that the $V_{ce} = 6$ V before even considering any input to the circuit. The advantage of mid-point biasing is that V_{ce} is safely half way between the cut-off and saturation boundaries, and can therefore vary almost equally up and down, and still remain in linear mode.

We have already considered a circuit with a d.c. bias (although it was not mentioned at the time) in the description of the CC circuit action shown in Figure 8.11(a). The trick is to make sure that the V_{in} to the circuit contains a d c. offset voltage. In Figure 8 11(a) the sine wave had a d.c. offset of 6 V, which meant that if the sine wave component of the signal had been reduced to zero, the BJT would still have been conducting. This d.c. offset is easy to add if the signal comes from a workshop waveform generator, but not quite so easy for other sources of voltage signal. How, therefore, do we add the d.c. bias? We will see, but first we need to understand one more term: 'quiescent'.

After a circuit has been biased and is conducting steady d.c. currents, nothing else happens until an a.c. signal is added which causes its currents and voltages to vary around the steady d.c. values. We could say that the circuit is idling or resting before being driven (a little like a car engine which is running but has not yet been put into gear). The circuit is said to be in its 'quiescent state', which means at rest. Therefore, we usually describe all such d.c. bias voltages and currents as quiescent voltages or quiescent currents. This adds a subscript q to all the terms, e.g. I_{cq} means quiescent collector current. So whenever we work out biasing values, we add the subscript q to demonstrate that we are talking about bias or quiescent values

Finally, setting the bias values to specific values is known as 'setting the Q point'. The Q point is defined as the quiescent voltage and current values of the BJT V_{ceq} and I_{cq} For example, if V_{ceq} = 3.6 V and I_{cq} = 10 mA, the Q point is (3 6 V, 10 mA).

There are four ways of introducing a d.c. bias into a circuit and in each case we call it a 'biasing network'. We will look at the standard common emitter circuit to explain each one.

8.10.3 Base biasing and Q point variation

We always try to obtain mid-point biasing, which means that we require V_{ceq} = V_{cc}/2. Refer to the circuit in Figure 8.14

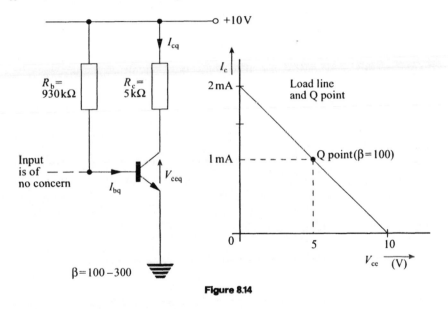

Figure 8.14

To find R_c in this circuit we want the BJT V_{ceq} to be 5 V, which means that R_c must also drop 5 V (to take up the remaining power supply volts). We usually have a choice here about what the collector current I_{cq} should be to cause the needed voltage drop. Let us say we choose I_{cq} = 1 mA Therefore

$$R_c = V_{Rc}/I_{cq}$$

$$= 5/1 \text{ mA}$$

$$= 5 \text{ k}\Omega$$

To find R_b and to generate $I_{cq} = 1$ mA, we need to calculate I_{bq} using the worst case β value. Therefore

$$I_{bq} = I_{cq}/\beta$$

$$= 1 \text{ mA}/100$$

$$= 0.01 \text{ mA}$$

I_{bq} comes from the power supply rail, down through R_b. We know that the voltage at the base end of R_b is 0.6 V. Therefore

$$R_b = (V_{cc} - V_{be})/I_{bq}$$

$$= 9.4 \text{ V}/0.01 \text{ mA}$$

$$= 940 \text{ k}\Omega$$

This has set up a Q point of (5 V, 1 mA).

To picture this, it would be useful to show the Q point on the d c. load line Therefore, we need to calculate the points at each intercept.

$V_{ce} \approx 0$ at the I_c intercept. Therefore

$$I_c = V_{cc}/R_c$$

$$= 10/5 \text{ k}\Omega$$

$$= 2 \text{ mA}$$

and $I_c = 0$ at the V_{ce} intercept. Therefore

$$V_{ce} = V_{cc}$$

$$= 10 \text{ V}$$

We can see from the load line in Figure 8.14 that the Q point is sitting safely at the half-way point along the V_{ce} axis.

We now add an a.c. signal to the base (coupled through a capacitor to ensure that the d.c. voltage levels are not affected). The a.c. signal then just makes all the voltage and current levels in the circuit rise and fall around the quiescent values. In other words, in a.c. operation the a c. voltages and currents all have a mean value equal to the bias or quiescent values. Look at Figure 8.15, where the voltage variations are shown With mid-point biasing the voltage and current variations are unclipped or undistorted. Clipping and distortion occurs, however, when the Q

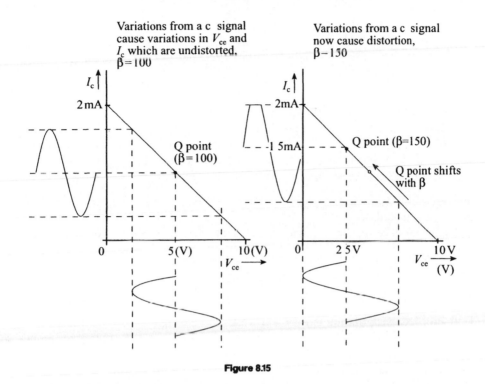

Figure 8.15

point moves away from this point What makes the Q point move away from the mid-point? Temperature changes and variations in β can shift the Q point

What happens if the β of the BJT happens to be nearer the top end of its range? This is where the problems creep in. Imagine replacing our BJT with one having $\beta = 150$ The calculations show that I_{bq} stays the same, but I_{cq} increases We have

$$I_{bq} = (V_{cc} - V_{be})/R_b$$

$$= (10 - 0 6)/940 \text{ k}\Omega$$

$$= 10 \,\mu\text{A (unchanged)}$$

Therefore

$$I_{cq} = I_{bq} \times \beta$$

$$= 10 \,\mu\text{A} \times 150$$

$$= 1 5 \text{ mA (increased)}$$

This increase in I_{cq} also creates a different V_{ceq}

$$V_{ceq} = V_{cc} - (I_{cq} \times R_c)$$

$$= 10 - 7.5$$

$$= 2.5 \text{ V}$$

Thus we have a new Q point at (2.5 V, 1.5 mA)

You can see the effect of this in Figure 8 15. The voltage and current variations cannot be reproduced without the BJT entering one of the cut-off or saturation modes Clipping of the waveform occurs as shown Base biasing is therefore not recommended because of its dependence on the value of β Why? Because.

1. We never know accurately the value of β (unless we measure it experimentally).
2. If we replace the BJT by one of the same type, its β will be different

Other biasing techniques have therefore been developed

8.10.4 Collector feedback biasing

Collector feedback biasing is easy and works quite well, even though the analysis displays a decidedly β-dependent quiescent point. Its success is based more on the fact that it uses negative feedback. You know instantly from this that the circuit must therefore be more stable than a circuit without negative feedback! Figure 8.16 shows a circuit with collector feedback biasing Although still β dependent, it has

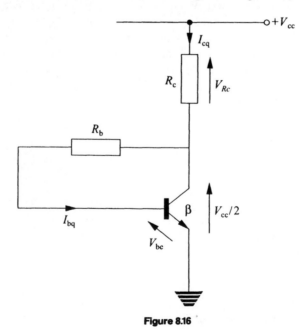

Figure 8.16

good immunity to β variations and the analysis is fairly straightforward. If we assume mid-point biasing, then $V_{ceq} = V_{cc}/2$, $V_{Rc} = V_{cc}/2$ and $I_{cq} = V_{Rc}/R_c = V_{cc}/(2 \times R_c)$. Therefore

$$I_{bq} = I_{cq}/\beta$$

$$= V_{cc}/(2 \times \beta \times R_c)$$

Since $R_b = [(V_{cc}/2) - V_{be}]/I_{bq}$, we have

$$R_b = [(V_{cc}/2) - V_{be}]/[V_{cc}/(2 \times \beta \times R_c)]$$

If $V_{be} \ll V_{cc}$, then we can neglect V_{be} and this expression becomes

$$R_b \approx (V_{cc}/2)/[V_{cc}/(2 \times \beta \times R_c)]$$

which cancels down to give

$$R_b \approx \beta R_c$$

This makes the calculation of R_b very simple once we have chosen an R_c value

The negative feedback action

But why is this any better than base biasing? It still has β in the calculations. The answer to this is in the rather clever negative feedback action which can be described as follows (try to follow the argument by referring to the circuit):

1. If the β value of the BJT is higher than that assumed in the calculation, I_{cq} rises. This, however, causes V_{ceq} to fall A fall in V_{ceq} means that I_{bq} must reduce. A reduction in I_{bq} means a reduction in I_{cq}, which pulls I_{cq} back down to its original value!
2. If the β value of the BJT is lower than that assumed in the calculation, I_{cq} falls. This, however, causes V_{ceq} to rise. A rise in V_{ceq} means that I_{bq} must increase. An increase in I_{bq} means an increase in I_{cq}, which pushes I_{cq} back up to its original value!

Clever old feedback

The Q point therefore remains pretty steady, even with changes in the value of β. The circuit uses current feedback, which means that the input resistance drops drastically. (This is a problem in its use as a voltage amplifier.)

8.10.5 Emitter biasing

Emitter biasing works excellently and is the sort of biasing that a BJT op-amp would use. The circuit is shown in Figure 8.17. Here the Q point is fixed and

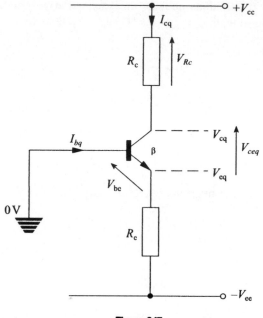

Figure 8.17

independent of β, which makes it a good bias network to choose. Notice that we now have an emitter resistor (R_e) and an extra power supply voltage ($-V_{ee}$). To find the Q point we first have to find I_{cq}

$$I_{cq} \approx I_{eq}$$

$$I_{eq} = (V_{eq} - V_{ee})/R_e$$

where $V_{eq} = -V_{be}$. Therefore

$$I_{cq} \approx (-V_{be} - V_{ee})/R_e$$

Next we find V_{ceq}

$$V_{ceq} = (V_{cq} - V_{eq})$$

where $V_{cq} = [(V_{cc} - (I_{cq} \times R_c))]$ and $V_{eq} = -V_{be}$. Therefore

$$V_{ceq} = [(V_{cc} - (I_{cq} \times R_c)) - (-V_{be})]$$
$$= [(V_{cc} - (I_{cq} \times R_c)) + V_{be}]$$

The result is interesting in that this Q point (V_{ceq}, I_{cq}) calculation does not use the β

value at all! There are four main points to note about emitter biasing:

1. It requires a bipolar power supply (could be a problem!).
2. If an input signal is applied to the base as it stands, the input is shorted to earth. To get around this, we insert a large resistor between the base and the earth and hope that the base current drawn by the BJT does not cause a significant voltage drop across the resistor (we still want the quiescent base voltage to be zero). The resistor added is called a 'ground return resistor' Does this ring a bell? (Real op-amps?)
3. R_e adds some negative feedback and therefore stabilizes the circuit.
4. The really good thing about this biasing method is that the quiescent emitter (and therefore the collector) current is set up exclusively by the voltage drop across R_e If we know V_{ee} and we know R_e, bingo, that is all we need!

EXAMPLE 8.4

Find the Q point of an emitter biased circuit with $R_c = 1\,k\Omega$, $R_e = 0\,5\,k\Omega$, $V_{cc} = 25\,V$ and $V_{ee} = -5\,V$ Draw the d c load line.

SOLUTION

We simply put these values into the Q point equations

$$I_{cq} \approx (-V_{be} - V_{ee})/R_e$$

Therefore

$$I_{cq} = [-0.6 - (-5)]/0.5 \text{ k}\Omega$$

$$= 4.4 \text{ V}/0.5 \text{ k}\Omega$$

$$= 8.8 \text{ mA}$$

$$V_{ceq} = [(V_{cc} - (I_{cq} \times R_c)) + V_{be}]$$

Therefore

$$V_{ceq} = [25 \text{ V} - (8.8 \text{ mA} \times 1 \text{ k}\Omega) + 0.6 \text{ V}]$$

$$= 16.8 \text{ V}$$

Therefore, the Q point is (16.8 V, 8.8 mA)

Finally, the load line intercepts can be found from $I_c = 0$ and $V_{ce} = 0$ When $I_c = 0$, the full power supply voltage is across V_{ce}. Therefore

$$V_{ce} = V_{cc} - (V_{ee})$$
$$= 25 - (-5)$$
$$= 30 \text{ V}$$

When $V_{ce} = 0$, the full power supply voltage is across $R_c + R_e$. Therefore

$$I_c = [V_{cc} - (V_{ee})]/(R_c + R_e)$$
$$= 30 \text{ V}/1.5 \text{ k}\Omega$$
$$\approx 20 \text{ mA}$$

The load line and Q point are shown in Figure 8.18

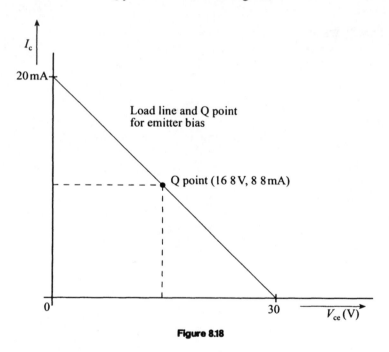

Figure 8.18

8.10.6 Voltage divider biasing

Voltage divider biasing is by far the most commonly known method and is used for quick BJT design. All the good things about emitter biasing are used here, except that we do not need a bipolar power supply. The voltage divider biasing method simply supplies a fixed positive voltage at the base of the BJT from a voltage

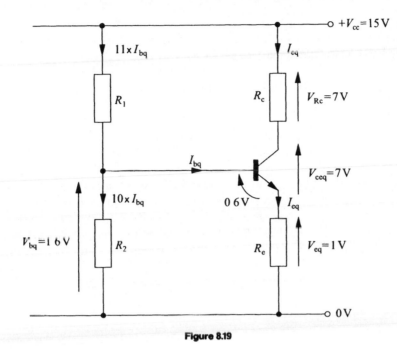

Figure 8.19

divider current The only difficulty is in making sure that the output voltage from the divider is not pulled down too much by the loading effect of the BJT circuit. The circuit is shown in Figure 8.19. It is useful to note the following points about voltage divider biasing:

1   It uses voltage negative feedback to stabilize the Q point. This is provided by R_e

2.   Since there is a +/−50 mV spread in V_{be} for a defined I_c value *and* a −2 1 mV/°C temperature dependency on V_{be}, it is best to allow V_{eq} to be much larger than this variation. We therefore make V_{eq} between 1 V and 3 V (the larger the better).

3   The fixed d.c. bias is generated by the voltage divider action of R_1 and R_2. Since I_{bq} is drawn from this, it is usual to allow the voltage divider current to be 10 × I_{bq} This makes sure that the BJT load current does not pull the bias voltage down too much. (We can also use a Thevenin equivalent circuit for the R_1, R_2 calculation. The R_{th} for the voltage divider is just $R_1 \parallel R_2$. V_{th} is the required V_{bq} value and the load on the Thevenin circuit is the input resistance of the BJT circuit R_{in}, where $R_{in} \approx \beta \times R_e$. Normally we are looking for good voltage transfer and therefore it is usual to choose $R_{th} \approx R_{in}/20$.)

4. R_e introduces negative feedback. If the temperature increases, I_{cq} rises and therefore V_{eq} rises Since V_{bq} is fixed (by the voltage divider circuit), this

causes V_{be} to fall, which of course causes I_{cq} to fall back to its original value! Its effect is to stabilize the circuit and fix the Q point: to stop it wandering up or down the load line

Biasing is determined by following a set of easy steps. To find the bias voltage and bias current:

1. Choose the required I_{cq} value for the circuit (this is up to you, the designer) and assume that I_{eq} is equal to it.
2. For a CE circuit, allow about 1 V to be dropped across R_e. Calculate R_e. (You know its voltage and current) For a CC circuit, allow half of V_{cc} to be dropped across R_e. Calculate R_e. (You know its voltage and current.)
3 For a CE circuit, the rest of the power supply must be dropped equally across the BJT and R_c. Calculate R_c. (You know its voltage and current.) For a CC circuit, the rest of the power supply must be dropped across the BJT.
4. The quiescent base voltage V_{bq} is simply the emitter voltage V_{eq} plus V_{be}. (Assume $V_{be} = 0.6\,V$). Calculate V_{bq}.
5. Find the lowest quoted β value for the BJT used and calculate I_{bq} (from $I_{bq} = I_{cq}/\beta$).
6. The voltage divider circuit of R_1 and R_2 must provide the required V_{bq} and I_{bq}.

In finding values of R_1 and R_2, it is important to note that the voltage divider output must not be affected by the varying load currents required by BJTs of the same type but differing β values! We solve this by the following steps:

1. Assume that the lower voltage divider resistor (R_2) carries ten times the I_{bq} value. Calculate R_2. (You know its voltage and current.)
2. The upper voltage divider resistor (R_1) therefore carries the R_2 current plus I_{bq}. This means that R_1 carries $(10 \times I_{bq}) + I_{bq} = 11 \times I_{bq}$. The voltage across R_2 is found from $(V_{cc} - V_{bq})$. Therefore, calculate R_1 (You know its voltage and current.)

The designer would now look at the a.c. analysis (which we cover in Chapter 9)

EXAMPLE 8.5
Design the voltage divider bias for a CE BJT circuit, where β is in the range 300–800 and the power supply is +15 V. You may assume an I_{cq} of 1 mA.

SOLUTION
Using the design steps just given.

1. $I_{eq} \approx I_{cq}$
    $= 1$ mA (given).
2. As the design is for a CE circuit, allow 1 V across R_e. Therefore

$$R_e = V_{eq}/I_{eq}$$

$$= 1 \text{ V}/1 \text{ mA}$$

$$= 1 \text{ k}\Omega$$

3   The supply voltage left is

$$(V_{cc} - V_{eq}) = 15 - 1$$

$$= 14 \text{ V}$$

Divide this equally across V_{ce} and R_c Therefore

$$V_{ce} = 14/2$$

$$= 7 \text{ V}$$

and

$$V_{Rc} = 7 \text{ V}$$

Therefore

$$R_c = V_{Rc}/I_{cq}$$

$$= 7 \text{ V}/1 \text{ mA}$$

$$= 7 \text{ k}\Omega$$

4   $V_{bq} = V_{eq} + V_{be}$
    $= 1 \text{ V} + 0.6 \text{ V}$
    $= 1.6 \text{ V}$
5   The lowest value of β is 300   Therefore

$$I_{bq} = I_{cq}/\beta$$

$$= 1 \text{ mA}/300$$

$$= 3.3 \,\mu\text{A}$$

6   The voltage divider must provide 1 6 V at 3.3 μA.
7   Assume that R_2 carries $10 \times I_{bq}$. Therefore

$$R_2 = V_{bq}/(10 \times I_{bq})$$
$$= 1.6 \text{ V}/33 \, \mu\text{A}$$
$$\approx 48.5 \text{ k}\Omega$$

8.   Now R_1 carries $11 \times I_{bq}$ and drops $(V_{cc} - V_{bq})$ V. Therefore

$$R_1 = (V_{cc} - V_{bq})/(11 \times I_{bq})$$
$$= (15 - 1.6) \text{ V}/(11 \times 3.3 \, \mu\text{A})$$
$$= 13.4 \text{ V}/36.3 \, \mu\text{A}$$
$$= 369 \text{ k}\Omega$$

Therefore, we have

$$R_1 = 369 \text{ k}\Omega$$
$$R_2 = 48.5 \text{ k}\Omega$$
$$R_c = 7 \text{ k}\Omega$$
$$R_e = 1 \text{ k}\Omega$$

For CC circuits the calculations differ only in that the voltages V_{ce} and V_e are both $V_{cc}/2$, which means that the voltage divider resistors tend to be almost equal.

Self-assessment questions

SAQ8.20 Is biasing concerned with a.c. or d c. values?

SAQ8.21 Design the voltage divider biasing for a CC circuit with a power supply of 9 V, $I_{cq} = 1$ mA and β in the range 100–500.

SAQ8.22 Draw a d.c load line for SAQ8.21.

SAQ8.23 Is the voltage divider bias method completely independent of the β value?

SAQ8.24 Design the voltage divider biasing for a CE circuit where $V_{cc} = 10$ V, $I_{cq} = 4$ mA and β is in the range 30–100.

SAQ8.25 Once biasing has been established, why do you think a capacitor is used between an external signal source and the BJT circuit?

SAQ8.26 The voltage divider biasing circuit requires good voltage transfer between the voltage divider and the BJT. Using the values given in Example 8.6, work out whether or not the values of R_1 and R_2 calculated would provide this good voltage transfer. (Hint: find the input resistance of the BJT circuit; use it as the load resistance on the voltage divider circuit; then find the Thevenin resistance of the voltage divider and apply the condition for reasonable voltage transfer (i e. $R_{th} < R_{in}/10$).)

Checklist

Can you now:
- state the relationship between emitter, collector and base currents?
- recognize which mode a BJT is working in?
- design a switch circuit to invert and amplify a square wave?
- design a BJT switch to give an alarm when a resistance goes too low?
- recommend a linear BJT circuit with a high voltage gain?
- recommend a linear BJT circuit with a high input resistance?
- explain why a common base circuit is good for high frequencies?
- explain four biasing networks and their advantages and disadvantages?
- design a CE voltage divider bias circuit?
- design a CC voltage divider bias circuit?

Tutorial questions

1   How is I_e related to I_b? Derive a simple expression relating them.
2   A BJT is used in a circuit where $V_{cc} = 12\,V$. Determine the BJT modes of operation if V_{ce} is measured as $0.2\,V$, $12\,V$, $11.5\,V$ and $3\,V$.
3   If $40 \leq h_{FE} \leq 200$ and $I_c = 1\,mA$, what is the maximum possible value of I_b?
4   If $1\,mA \leq I_b \leq 10\,mA$ and I_c needs to be $150\,mA$, what is the minimum β requirement?
5   Design a BJT switch circuit to switch an $18\,V$, $2\,W$ lamp on and off from a 0–$12\,V$ input square wave. Specify your own β and $I_{c(max)}$ values for the BJT and the required base current.
6   A certain BJT has $I_e = 1\,mA$. If the V_{be} value increases by $54\,mV$, what is the new I_c value?
7   A BJT is set up with $I_e = 25\,mA$ and $I_b = 100\,\mu A$. What are the values of h_{FE} and I_b?
8   Why does a CE circuit usually have a resistor between the emitter and earth? What is this circuit amendment called?
9   Explain why biasing is usually necessary for linear BJT operation
10   What is the meaning of mid-point biasing?

11   Referring to Figure 8.14, design a base bias circuit for $V_{cc} = 24\,V$ and $I_{cq} = 15\,mA$. Assume a typical β value of 100.
12   Draw the load line for the answer to question 11. Show also the movement of the Q point if the BJT is replaced by one with $\beta = 150$.
13   Name the four main biasing techniques used for BJT circuits.
14   Design the voltage divider biasing for a CC circuit whose power supply is $18\,V$, $I_{cq} = 10\,mA$ and β is in the range 150–500.
15   Draw the d.c. load line for the circuit in question 14.
16   Design the voltage divider biasing for a CE circuit whose power supply is $12\,V$, $I_{cq} = 5\,mA$ and β is in the range 40–200.
17   Draw the d.c. load line for the circuit in question 16.
18   Find the TEC for the voltage divider part of the circuit in question 14 and, by using the V_{bq} and I_{bq} values needed by the BJT, determine whether or not the biasing circuit is a reasonable voltage source. How would you change the biasing design if you needed to improve the voltage transfer between the divider circuit and the BJT?
19   Design the voltage divider biasing for a CE circuit whose power supply is $5\,V$, $I_{cq} = 0.5\,mA$ and β is in the range 300–500.
20   Determine the new Q point position in question 19 if the BJT were replaced with one with $\beta = 400$.

Further reading

Horowitz P & Hill W. (1987). *The art of electronics*, sections 2 01–3, 2 07, 2 11–12
    Cambridge University Press, Cambridge, England
Millman J & Halkias C (1988). *Integrated electronics*, chap. 5 McGraw-Hill
Ritchie G J (1987) *Transistor circuit techniques*, pp. 12–19, 28–36, 102–4 Van Nostrand
    Reinhold, Wokingham, England.

CHAPTER 9

BJT a.c. analysis

Aims and objectives

This chapter aims to teach you the skills required to design a voltage amplifier and a voltage buffer circuit to a given specification. You will be shown how to model the BJT for a.c. operation and how to predict the basic interface characteristics of the circuits.

After studying this chapter you will be able to:

- *understand how to apply and use the r_e a.c. model for the BJT*
- *draw a.c. and d.c. equivalent circuits*
- *calculate approximate input and output resistances for CC and CE circuits*
- *design a voltage amplifier with a specific gain value*
- *use coupling and bypass capacitors to tailor frequency response and voltage gain of a circuit*

We have considered BJT operation and the d.c. biasing of a circuit in some detail. We will now investigate how the BJT operates under a.c. conditions. Most analog BJT circuits will be dealing with a.c. signals such as sine waves, so an understanding of the a.c. analysis of a circuit is vital to the designer. In most cases (as usual) the a.c. responses can be approximated to give us design rules that are easier to use.

9.1 Naming a.c. quantities

It is traditional when considering a.c. quantities to use lower case letters like i_c for a.c. collector current and r_{in} for a.c. input resistance. We will therefore use this type of definition throughout this chapter There have been occasions in earlier chapters where the differences in a.c. and d.c. quantities were not very important and therefore were not shown separately.

233

9.2 Intrinsic emitter resistance and the BJT model

In Chapter 8 we discussed briefly the fact that the V_{be} and I_e of the BJT are related by the equation

$$I_e \approx I_s[\exp(V_{be}/K)]$$

where K is a constant of about 0.026 V at room temperature. This is often called the Shockley equation after William Shockley, one of the inventors of the BJT back in 1951. Since $I_c \approx I_e$, we can say

$$I_c \approx I_s[\exp(V_{be}/K)]$$

Just as in the case of the diode curve, this equation is a mathematical representation of the I_c/V_{be} curve for a BJT shown in Figure 9.1. The gradient of this curve is $\Delta I_c / \Delta V_{be}$. Its reciprocal $\Delta V_{be}/\Delta I_c$ is the a.c. resistance (r_e) of the BJT over a small change in V_{be}. Using calculus we can find a value of this a.c. resistance (r_e), as the small change tends towards zero. In other words, the differential dV_{be}/dI_e will give us the r_e of the BJT at the point where $I_c = I_{cq}$. We therefore differentiate the Shockley equation with respect to V_{be} and find its reciprocal. (You can try this in a SAQ later if you wish.) The result is very simple: the BJT a.c. resistance is given by

$$r_e \approx K/I_{cq}$$

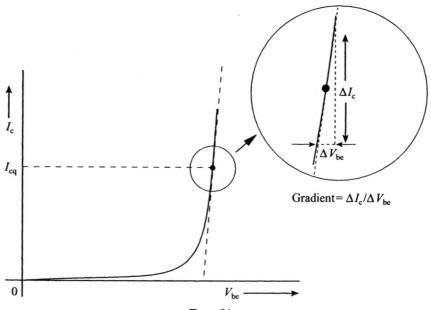

Figure 9.1

where $K = 0.026$ V and I_{cq} is the steady d.c collector current. To make things easier and because I_{cq} is usually quoted in mA, we usually multiply the top and bottom of this equation by 1000, to give

$$r_e \approx 26/I_{cq}$$

where I_{cq} is in milliamps and r_e is in ohms.

The r_e BJT model

The common name of r_e is 'intrinsic emitter resistance'. This merely means that the a c. resistance value is part of, or inside, the BJT. It appears *only* when a c. signals are present and it looks like a small resistor in the emitter leg of the BJT. We can model this effect as shown in Figure 9.2(a). This symbol is, however, a little clumsy to draw and it is better to use a more professional-looking model which includes the a.c current flow. What else can we include in our model? Well, we already know that an a c. variation in V_{be} results in a large a c. variation in I_c We can therefore show the resulting I_c variation as a small a c. current generator in the collector i_c. This new model is shown in Figure 9 2(b) and is known as the 'r_e BJT model'. The expression for r_e allows us to approximate its value easily when we know the collector bias current value.

EXAMPLE 9.1

Determine the intrinsic emitter resistance of a BJT biased with a Q point of (10 5 V, 4 mA).

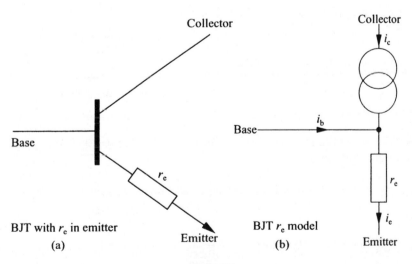

BJT with r_e in emitter
(a)

BJT r_e model
(b)

Figure 9.2

SOLUTION

We ignore the V_{ceq} value given and use only the I_{cq} value. Therefore

$$I_{cq} = 4 \text{ mA (given)}$$

The intrinsic emitter resistance is

$$r_e \approx 26/I_{cq}$$
$$= 26/4$$
$$= 6.5 \, \Omega$$

EXAMPLE 9.2

Determine the intrinsic emitter resistance of a BJT biased with a Q point of (2.8 V, 0.5 mA).

SOLUTION

$$I_{cq} = 0.5 \text{ mA (given)}$$

Therefore the intrinsic emitter resistance is

$$r_e \approx 26/I_{cq}$$
$$= 26/0.5$$
$$= 52 \, \Omega$$

Notice how r_e increases as I_{cq} decreases

Self-assessment questions

SAQ9.1 Find the intrinsic emitter resistance of a BJT whose Q point is (3.6 V, 1.5 mA).

SAQ9.2 The r_e value for a certain BJT is 10 Ω. What is the quiescent collector current value?

SAQ9.3 Determine the r_e values for a BJT whose I_{cq} values are:
1.   2 A;
2.   50 μA.

SAQ9.4 What is the approximate Q point of a mid-point biased CC circuit with a 12 V power supply and a BJT intrinsic emitter resistance of 52 Ω?

SAQ9.5 Try to derive the $r_e \approx 26/I_{cq}$ expression by differentiating the Shockley equation. (Hint: differentiate with respect to V_{be} and then substitute the original equation into the differential.)

9.3 Using the r_e model to find the BJT a.c. input resistance

This BJT a.c model allows us to do some simple analysis on a BJT circuit. One of the parameters of interest is the a c. input resistance (r_{in}) Figure 9.3 shows the CC configuration (no bias resistors are shown: we will consider their effect later) with the circuit redrawn to show our r_e a.c model. Using the general rule that

input resistance = input voltage/input current

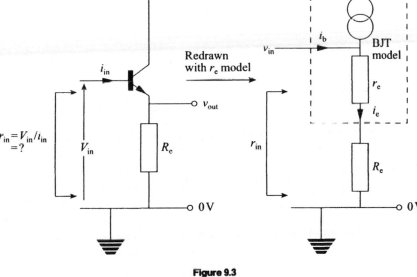

Figure 9.3

Therefore

$$r_{in} = v_{in}/i_{in}$$

In this circuit, v_{in} drives the input current i_b into the BJT. Therefore

$$i_{in} = i_b$$

and therefore

$$r_{in} = v_{in}/i_b$$

i_b causes i_e to flow down through r_e and R_e to earth. The sum of the voltage drops across r_e and R_e is equal to v_{in}. Therefore, we can say

$$v_{in} = i_e \times (r_e + R_e)$$

However, we know that

$$i_e = i_b + i_c$$
$$= i_b + (\beta \times i_b)$$
$$= i_b(1 + \beta)$$

Therefore

$$v_{in} = i_b(1 + \beta) \times (r_e + R_e)$$

Finally therefore

$$r_{in} = [i_b(1 + \beta) \times (r_e + R_e)]/i_b$$

Cancelling i_b gives the a c. input resistance as

$$r_{in} = (\beta + 1) \times (r_e + R_e)$$

Since $\beta \ll 1$, this expression is always approximated to

$$r_{in} \approx \beta(r_e + R_e)$$

and if

$$R_e \gg r_e$$

then

$$r_{in} \approx \beta \times R_e$$

In some circuits R_e is omitted. This removes its feedback action and leaves us with

$$r_{in} = \beta \times r_e$$

The BJT appears to make any emitter resistance look β times larger than its actual value when the circuit is viewed from the input. The r_{in} for an emitter follower is β times the emitter resistance. The emitter resistance can be taken as R_e

or $R_e + r_e$ (for greater accuracy). If there is no external emitter resistor, then it is simply $\beta \times r_e$.

EXAMPLE 9.3

Find r_{in} for Figure 9.4(a)–(c). Assume all circuits are at room temperature.

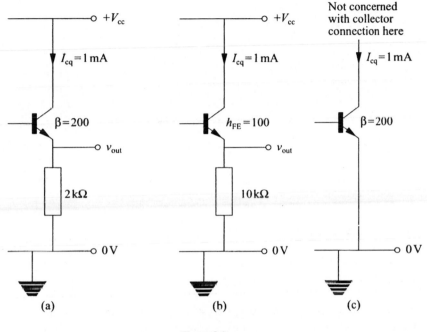

Figure 9.4

SOLUTION

1. For Figure 9 4(a), using $r_{in} = (\beta + 1) \times (r_e + R_e)$ and $r_e = 26/I_{cq}$, we have $\beta = 200$, $r_e = 26/1$ mA $= 26\,\Omega$ and $R_e = 2\,k\Omega$ Therefore

$$r_{in} = 201 \times 2026\ \Omega$$

$$= 407.2\ k\Omega$$

Using the approximation $r_{in} \approx \beta \times R_e$, we have

$$r_{in} \approx 200 \times 2\ k\Omega$$

$$\approx 400\ k\Omega$$

2 Using the same method (remember $h_{FE} = \beta$) for Figure 9 4(b), we have

$$r_{in} = (100 + 1) \times ([26/1]\Omega + 10 \text{ k}\Omega)$$
$$= 1.01 \text{ M}\Omega$$

or

$$r_{in} \approx 100 \times 10 \text{ k}\Omega$$
$$\approx 1 \text{ M}\Omega$$

3. For Figure 9.4(c), $R_e = 0$. Therefore

$$r_{in} = (200 + 1) \times (0 + [26/1]\Omega)$$
$$= 5.23 \text{ k}\Omega$$

or

$$r_{in} \approx 200 \times [26/1]\Omega$$
$$\approx 5.2 \text{ k}\Omega$$

Note that the approximations are quite adequate for most calculations.

Self-assessment questions

SAQ9.6 What is the input resistance of a BJT CC circuit whose β is 200, Q point is (10 V, 10 mA) and R_e is 1 kΩ?

SAQ9.7 In a certain CC circuit $R_e = 15$ kΩ and I_{cq} is a few milliamps. Can we neglect r_e in the r_{in} calculation?

SAQ9.8 If we wanted a higher input resistance, would we increase or decrease the quiescent collector current?

SAQ9.9 If we have a circuit where the emitter is grounded, what happens to the r_{in} if the collector current increases?

SAQ9.10 In a grounded emitter circuit, the r_{in} appears to vary with the size of the input signal. What would be a good way of stabilizing the input resistance?

9.4 A.c. and d.c. equivalent circuits

We now come to a very important step in our journey through electronics. Being able to use models and draw equivalent circuits is our way of making life easier.

Understanding and designing circuits can be made much easier if we can replace real devices by simplified mathematical models of the same thing.

There are many levels of model and equivalence we can choose. Usually, more complex models yield more exact answers but take longer to compute and are more difficult to understand. At this level we are aiming to grasp an understanding of circuit action and to generate approximate but usable answers. More rigorous and complex device analysis can be investigated later (if necessary), but it will usually be done for you by a software package!

9.4.1 Equivalent circuits of a coupling or bypass capacitor

Why capacitors have the names 'coupling' and 'bypass' will become more clear in a moment. First, let us see how a capacitor behaves when driven with an a.c. signal. Remember that for a.c signals the capacitor has a frequency-dependent reactance value X_c, where

$$X_c = 1/j\omega C$$

$\omega = 2\pi f$ and j indicates a phase change when currents flow. We will ignore the phase-changing property as this is not relevant to our discussion here. The expression shows a reactance value inversely proportional to frequency, i.e. as f rises, X_c falls, and vice versa. If we consider each end of the frequency range, this leads us to two conclusions:

1 As f reduces to d.c., $f = 0$ Hz and we have

$$X_c = 1/0 = \infty \ \Omega.$$

2 As f increases towards infinity, $f \to$ infinity and we have

$$X_c \to 1/\infty \to 0 \ \Omega.$$

What use are these? Well, we often consider d.c conditions in isolation Take the d.c BJT biasing calculations, for example. At d.c the capacitance has infinite reactance, which effectively makes it an open circuit At the other end of the scale, capacitors have low reactance values at higher frequencies. In fact, the designer usually chooses values of C which make X_c negligible at all the frequencies likely to be encountered for the circuit. A good approximation for a negligible X_c value is a short circuit. Look at Figure 9.5 to see these ideas.

What we are really saying is that if we insert a capacitor in a circuit, it allows a.c signals to pass through without attenuation (reduction in voltage), but it blocks the path of any d.c. levels, allowing different d.c. voltages to remain across its ends. This gives us two rules for capacitors in equivalent circuits:

1 capacitors in d c. equivalent circuits replace all capacitors by an open circuit,

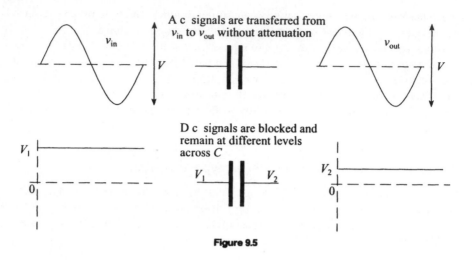

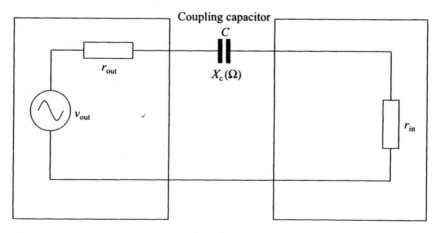

Figure 9.5

2. capacitors in a.c. equivalent circuits. replace all capacitors by a short circuit.

Coupling capacitors

Capacitors are used in this way to 'couple' a.c. circuits together. This means that the a c. signal can be transferred from the source circuit to the destination circuit without any steady d.c. bias conditions being affected: hence the name 'coupling capacitors'. Figure 9.6 illustrates this.

To ensure that the capacitor will transfer most of the a c. signal voltage with as little attenuation as possible, we make sure that its X_c is small compared with the other resistances in the circuit. Small usually means 10% or less. Since X_c is at its

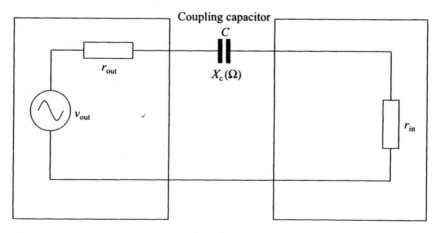

Figure 9.6

largest at lower frequencies, we always choose to design at the minimum frequency (f_{min}) specified for a circuit. This is another example of worst case design. Referring to Figure 9.6, we would choose

$$X_c = (r_{in} + r_{out})/10$$

(For some designs this can generate enormous C values, so designers opt for a more realistic $X_c = (r_{in} + r_{out})$ value.)

EXAMPLE 9.4

Find a suitable input coupling capacitor for an amplifier with a frequency response of 50 Hz to 15 kHz and an $r_{in} = 50\,k\Omega$, driven from a voltage source with $r_{out} = 50\,k\Omega$.

SOLUTION

Using $X_c = (r_{in} + r_{out})/10$, we have

$$r_{in} = 50\ k\Omega$$

$$r_{out} = 50\ k\Omega$$

$$f_{min} = 50\ Hz$$

Therefore

$$X_c = 1/(2 \times \pi \times f_{min} \times C)$$

$$= (r_{in} + r_{out})/10$$

Therefore

$$C = 10/[2 \times \pi \times f_{min} \times (r_{in} + r_{out})]$$

$$\approx 0.3\ \mu F$$

Self-assessment questions

SAQ9.11 Find a suitable input coupling capacitor for an amplifier with a frequency response of 20 Hz to 20 kHz and $r_{in} = 10\,k\Omega$, driven from a voltage source with $r_{out} = 10\,k\Omega$

SAQ9.12 Determine a suitable output coupling capacitor for an amplifier with a frequency response of 2 kHz to 10 kHz and $r_{out} = 10\,k\Omega$, to drive a circuit with $r_{in} = 500\,k\Omega$.

Bypass capacitors

We often see capacitors used to transfer a.c. signals across a component (or series of components) and therefore bypassing the component's action for a.c. but not d.c. An a.c. earth is an example of this In Figure 9.7 we can see that the bypass capacitor C transfers a.c signals directly to earth, bypassing R For a c., R is short-circuited and therefore its action is ignored, whereas for d.c. R comes into play and C is considered as an open circuit.

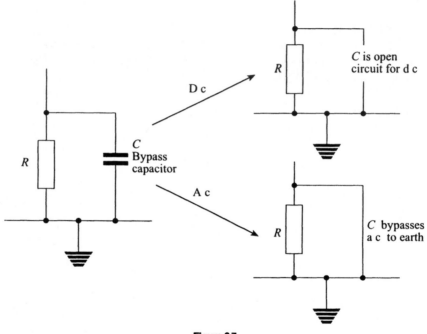

Figure 9.7

As in the case of coupling capacitors, we design to ensure that X_c is 10% or less than the R being bypassed. Therefore, we use

$$X_c = R/10$$

EXAMPLE 9.5

Choose a capacitor to create an a.c. bypass across a 1 kΩ resistance at all frequencies in the range 25 Hz to 10 kHz.

SOLUTION

Using $X_c = R/10$, where $R = 1$ kΩ and $f_{min} = 25$ Hz, we have

$$1/(2 \times \pi \times f_{min} \times C) = R/10$$

Therefore

$$C - 10/(2 \times \pi \times f_{min} \times R)$$

$$= 63\,\mu F$$

9.4.2 Equivalent circuit of a power supply

All active circuits require some form of energy source. This energy is usually supplied by the addition of an external d.c. power source to the circuit. The nature of the power source is that it behaves as a good source of voltage or current, which requires it to give a steady output irrespective of the load on it. We do not want a power supply the output of which varies with variations in load current drawn from it

In terms, therefore, of a c. and d.c quantities, we can say that the d.c quantity (or component) is constant and finite, and the a.c. quantity (or component) is non-existent or zero This means that a power supply will be shown in a d c. equivalent circuit as a specific voltage, but for a.c equivalent circuits it must be shown as being equal to zero. The only way an a.c. zero voltage can be modelled is by a short circuit. Therefore, the rules are

- For power supplies in d c equivalent circuits. show all power supply voltages and currents in full

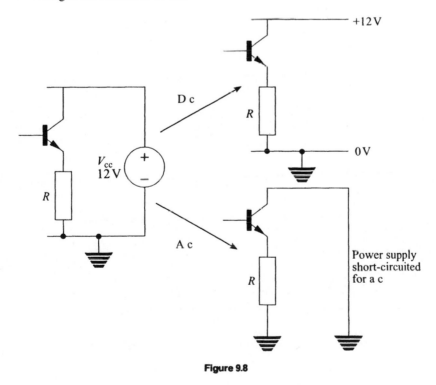

Figure 9.8

- For power supplies in a.c. equivalent circuits· replace all power supply voltages with a short circuit.

(See Figure 9 8.)

9.4.3 Drawing full a.c. and d.c. equivalent circuits

We have now reached a point where we can treat circuits from two different viewpoints We can separate and analyse a circuit for the d.c. conditions (like biasing and d.c. voltage gain) and a.c. conditions (like a.c. voltage gain and a.c. resistances). We must remember to use our BJT r_e model when required.

Let us now consider drawing the CE and CC BJT circuits as they would be used in a.c. applications. From these circuits we will then obtain the d.c. and a.c. equivalent circuits and use them to find out more about the circuit parameters.

9.5 The CC circuit

The a.c. emitter follower (CC) circuit, when used for a.c signals, looks like the circuit in Figure 9.9. Note the coupling capacitors linking the input and output to other circuits. The full a c. equivalent is shown in Figure 9.10 and its d.c. equivalent is shown in Figure 9.11

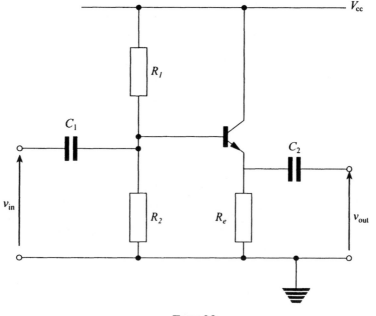

Figure 9.9

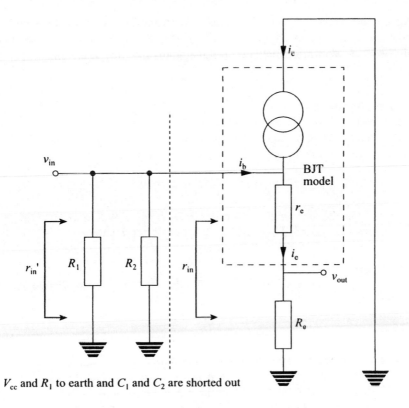

V_{cc} and R_1 to earth and C_1 and C_2 are shorted out

Figure 9.10

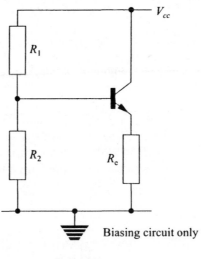

Biasing circuit only

Figure 9.11

We can determine the d.c bias conditions from the d.c. equivalent circuit in exactly the same way as is discussed in Chapter 8, and from the a c. equivalent circuit we can determine the a.c. voltage gain and the a.c. input and output resistances of the CC.

The a.c. voltage gain

We have assumed up to now that the A_v of the CC is unity and that $v_{in} = v_{out}$ The equivalent circuit demonstrates that the voltage drop across r_e caused by i_e will reduce v_{out} slightly. Therefore, $v_{out} = (v_{in} - (i_e \times r_e))$ and hence

$$A_v = (v_{in} - (i_e \times r_e))/v_{in}$$

This means that A_v is slightly reduced. Since $i_e \times r_e$ will only ever be a few millivolts, we are still happy to use $A_v \approx 1$ for all applications.

A.c. input resistance

Now look closely at the equivalent circuit We have already shown that the r_{in} value of the BJT is

$$r_{in} \approx \beta(R_e + r_e)$$

This is shown in Figure 9.10 to the right of the dotted line. What about the total a c. input resistance of the circuit (r_{in}')? Is it the same as r_{in}? The answer is no; it is rather less than r_{in}. When the a.c. circuit is drawn, the short-circuited power supply to earth causes the upper voltage divider resistance R_1 to be connected across the input, in parallel with R_2. The total r_{in}' value then becomes

$$r_{in}' = R_1 \parallel R_2 \parallel r_{in}$$

For most circuits this value is dominated by the biasing resistors R_1 and R_2. This is a shame, however, since the purpose of the emitter follower circuit is to present a high input resistance to a drive circuit and, although r_{in} can be made very large, the biasing resistors pull this value down by parallel resistance action.

The bias resistors degrade the CC's performance as a buffer circuit and as an a c. current amplifier. Its a c. input resistance reduces and therefore the a.c input current requirement increases. The only way to improve this is to apply bootstrapping techniques (refer to more detailed textbooks for this), to use emitter biasing or to arrange for a d c. offset to be present on the input signal.

EXAMPLE 9.6

Find the circuit input resistance of a voltage divider biased emitter follower circuit, with a Q point of (5 V, 1 mA) whose bias resistors are 1 kΩ and 1 2 kΩ with $R_e = 5$ kΩ and h_{FE} in the range 200–500

SOLUTION

The worst case design tells us to use $h_{FE} = 200$. Using $r_{in}' = R_1 \parallel R_2 \parallel r_{in}$, where $r_{in} \approx \beta(R_e + r_e)$ and $r_e - 26/I_{cq}$, we have

$$r_{in}' = 1 \text{ k}\Omega \parallel 1.2 \text{ k}\Omega \parallel 200[5 \text{ k}\Omega + (26/1\Omega)]$$

$$\approx 545 \, \Omega$$

Not very large! Note the dominating effect of R_1 and R_2

A.c. output resistance

Finding the output resistance value (r_{out}) is slightly more difficult, but is made easier if we make a logical deduction about the BJT action We noted earlier that the BJT appears to *amplify* the size of any emitter resistance by β when viewed from the input, i e $r_{in} \approx \beta(R_e + r_e)$. The converse of this also turns out to be true. The BJT appears to *reduce* the size of any resistance on the base side of the circuit by β when viewed from the output!

With reference to Figure 9.12, we can combine the bias resistances into R_b and include any source resistance R_s Using our new deduction, we can see that·

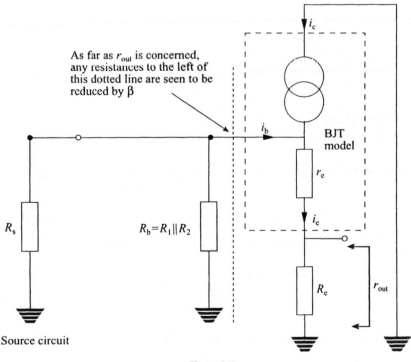

As far as r_{out} is concerned, any resistances to the left of this dotted line are seen to be reduced by β

i_c

BJT model

r_e

i_c

i_b

R_s

$R_b = R_1 \parallel R_2$

R_e

r_{out}

Source circuit

Figure 9.12

1   $R_b \parallel R_s$ will be reduced by β, since they are all on the input circuit side of the BJT. This will give us a value of $[(R_b \parallel R_s)/\beta]$.

2. Since R_e is directly across the output, it will remain in parallel with any other resistances.

3. r_e appears in series with $[(R_b \parallel R_s)/\beta]$.

Therefore, the full r_{out} expression will be

$$r_{out} \approx R_e \parallel r_e + [(R_b \parallel R_s)/\beta]$$

This looks difficult, but we can simplify it. Since β is typically about 100

$$[(R_b \parallel R_s)/\beta] \to 0$$

(especially if R_s is small!) and therefore

$$r_e + [(R_b \parallel R_s)/\beta] \to r_e$$

This leaves us with
$$r_{out} \approx R_e \parallel r_e$$

and since $r_e \ll R_e$, the expression approximates to

$$r_{out} \approx r_e$$

This gives us a good indication that r_{out} is quite small and therefore makes the emitter follower output a good source of voltage. Which is what we have said all along!

EXAMPLE 9.7
Find the approximate r_{out} for the circuit described in Example 9.6.

SOLUTION
We use $r_{out} \approx R_e \parallel \{r_e + [(R_b \parallel R_s)/\beta]\}$. We are not given R_s and therefore we must leave it out of the expression, with the knowledge that the r_{out} value will decrease as soon as an input source is connected to the circuit. We get

$$r_{out} \approx 5 \text{ k}\Omega \parallel \{(26/1\Omega) + [(1 \text{ k}\Omega \parallel 1.2 \text{ k}\Omega)/200]\}$$

$$\approx 29 \, \Omega (28.56 \, \Omega)$$

If we use the approximation $r_{out} \approx r_e$, we get

$$r_{out} \approx 26 \, \Omega$$

Note that this approximation is quite adequate

Self-assessment questions

SAQ9.13 Find the circuit input and output resistances for a CC circuit with $175 \leq h_{FE} \leq 500$, $R_1 = 10\,k\Omega$, $R_2 = 10\,k\Omega$, $R_e = 1\,k\Omega$ and a Q point of (5 V, 4 4 mA)

SAQ9.14 What is the difference between r_{in} and r_{in}' in SAQ9 13? What difference do you think this makes to the circuit current gain value?

9.6 The CE circuit

Analysis of the a.c parameters of the CE is easy, since much of the work is a duplication of the CC analysis. There are some variations, however, so we still need to consider it The a c CE circuit, when used for a c signals, looks like the circuit in Figure 9 13 Again, we have coupling capacitors linking the input and output to other circuits.

Now let us look at its full a.c equivalent shown in Figure 9 14 and its d c equivalent in Figure 9 15 We can determine the d c bias conditions from the d.c equivalent circuit in exactly the same way as is discussed in Chapter 8 The a c equivalent circuit is slightly different from the CC in that we now have R_c connecting the BJT collector to earth via the shorted power supply As before, we

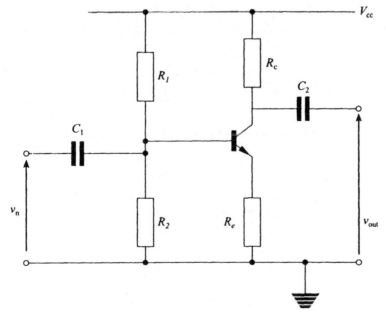

Figure 9 13

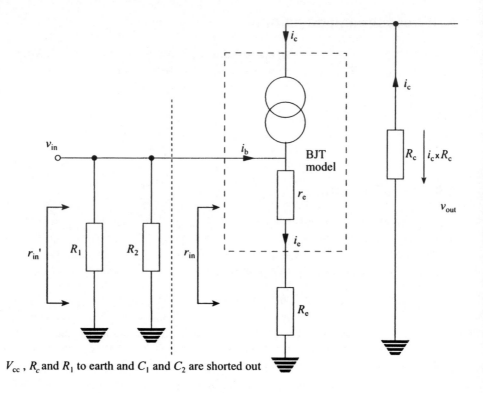

V_{cc}, R_c and R_1 to earth and C_1 and C_2 are shorted out

Figure 9.14

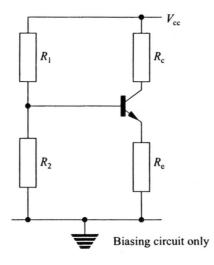

Biasing circuit only

Figure 9.15

can use this circuit to determine the a.c. voltage gain and the a.c. input and output resistances of the CE

The a.c. voltage gain

We are interested in finding expressions for v_{in} and v_{out} to allow us to set up a voltage gain equation where $A_v = v_{out}/v_{in}$. With reference to Figure 9.14, we can see that

$$v_{in} = i_e(r_e + R_e)$$

The expression for v_{out} needs careful consideration Notice that v_{out} is created by i_c flowing through R_c, but the direction of $i_c \times R_c$ is opposite to that of v_{out}. Therefore

$$v_{out} = -i_c \times R_c$$

This gives us

$$A_v = (-i_c \times R_c)/[i_e(r_e + R_e)]$$

If we assume $i_e \approx i_c$ (as usual), we have

$$A_v = (-i_c \times R_c)/[i_c(r_e + R_e)]$$

Cancelling i_c gives

$$A_v = -R_c/(r_e + R_e)$$

This is merely a ratio of two resistances and does not involve BJT parameters at all Note that the minus sign indicates the inverting action of the amplifier. We now have an easy way to deduce the A_v of a CE circuit Very often $r_e \ll R_e$ and we can simplify this expression even further to

$$A_v = -R_c/R_e$$

It could not be easier!

EXAMPLE 9.8

Determine the voltage gain of a CE circuit with $R_c = 10\,k\Omega$, $R_e = 1\,k\Omega$ and a Q point of (5 V, 1 mA)

SOLUTION

Using $A_v = -R_c/(r_e + R_e)$, we have

$$A_v = -10\ k\Omega/[(26/1)\Omega + 1\ k\Omega]$$

$$= -9.75$$

or more simply

$$A_v = -R_c/R_e$$

$$= -10 \text{ k}\Omega/1 \text{ k}\Omega$$

$$= -10$$

Self-assessment questions

SAQ9.15 Find the A_v of a CE circuit with $R_c = 20 \text{ k}\Omega$, $R_e = 2\ 5 \text{ k}\Omega$ and $I_{cq} = 500\ \mu\text{A}$

SAQ9.16 Would we increase or decrease I_{cq} to increase A_v?

SAQ9.17 If a CE circuit with $R_c = 10 \text{ k}\Omega$ and $I_{cq} = 1$ mA is used with a grounded emitter, find A_v

SAQ9.18 What will happen to A_v in SAQ 9 17 when the collector current rises?

SAQ9.19 A grounded emitter CE has an A_v which varies with input signal. Can you explain why and suggest an improvement to the circuit?

SAQ9.20 A CE with $A_v = -20$ has a 100 mV$_{rms}$ input signal. Sketch the input and output waveforms of the circuit.

A.c. input resistance

The r_{in}' value for the CE is identical to that for the CC. The total r_{in}' value is

$$r_{in}' \approx R_1 \parallel R_2 \parallel \beta(R_e + r_e)$$

As in the case of the CC circuit, its value is dominated by the biasing resistors R_1 and R_2.

EXAMPLE 9.9

Find the circuit input resistance of a voltage divider biased common emitter circuit, with a Q point of (6 V, 4 mA) whose bias resistors are 5 kΩ and 13 kΩ when $R_e = 0\ 75$ kΩ, $R_c = 1.5$ kΩ and h_{FE} is in the range 50–300.

SOLUTION

The worst case design tells us to use $h_{FE} = 50$ Using $r_{in}' = R_1 \parallel R_2 \parallel r_{in}$, where $r_{in} \approx \beta(R_e + r_e)$ and $r_e = 26/I_{cq}$, we have

$$r_{in}' = 5 \text{ k}\Omega \parallel 13 \text{ k}\Omega \parallel \{50[0.75 \text{ k}\Omega + (26/4)\Omega]\}$$

$$\approx 3.3 \text{ k}\Omega$$

Self-assessment question

SAQ9.21 In Example 9.9, what would happen to r_{in}' if we replaced the BJT with one with $\beta = 300$?

A.c. output resistance

Finding the output resistance value of the CE requires a little more consideration. Referring to Figure 9.14, we can see that R_c appears directly across the circuit output and therefore R_c will be in parallel with the output resistance of the rest of the circuitry. So what is the output resistance of the rest of the circuit? The starting point is that the BJT collector current source has a very high internal resistance (quoted in data sheets as $1/h_{oe}$), typically >200 kΩ. We can therefore redraw this circuit, replacing the current source with its (Norton) resistance. While we are about it, let us replace all the resistances on the input of the circuit with a single equivalent resistance R_b. This new equivalent circuit is shown in Figure 9.16.

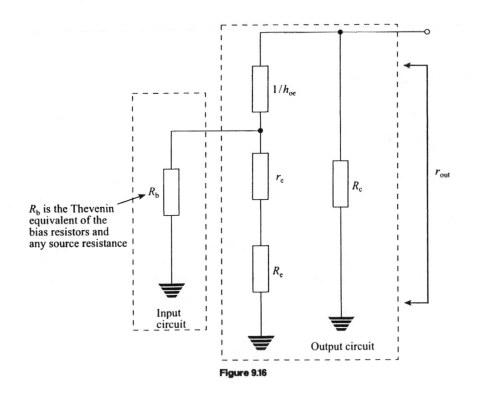

R_b is the Thevenin equivalent of the bias resistors and any source resistance

Figure 9.16

Remember from earlier that R_b is on the input side of the circuit and therefore appears to be reduced by β as far as the output is concerned and becomes (R_b/β). The reduced R_b/β resistance is in parallel with the series emitter resistances $(R_e + r_e)$, giving a combined value of $(R_b/\beta) \parallel (R_e + r_e)$, which itself is in series with the current source resistance $1/h_{oe}$, giving

$$[(1/h_{oe}) + (R_b/\beta) \parallel (R_e + r_e)]$$

Finally, R_c is in parallel with all this and our expression becomes

$$r_{out} = R_c \parallel [(1/h_{oe}) + (R_b/\beta) \parallel (R_e + r_e)]$$

Not very nice! Now let us simplify this. Since $[(R_b/\beta) \parallel (R_e + r_e)] \ll 1/h_{oe}$, we can neglect it to leave only $1/h_{oe}$, giving

$$r_{out} \approx R_c \parallel (1/h_{oe})$$

Usually, $R_c < 1/h_{oe}$. Therefore, we end up with the expression that everybody uses

$$r_{out} \approx R_c$$

All we need to remember, therefore, is that the output resistance of a CE circuit is approximately equal to the collector resistor!

EXAMPLE 9.10
Find the circuit output resistance of the circuit described in Example 9 9.

SOLUTION
We use $r_{out} \approx R_c$. We simply say that since $R_c = 1.5\,k\Omega$

$$r_{out} \approx 1.5\ k\Omega$$

Using h parameters
You may have wondered what h_{oe} and perhaps h_{FE} mean. Well, traditionally the BJT a.c. model was called the 'h parameter' model and was made up of parameter values which can be measured experimentally in the laboratory. If the measurements are made in the common emitter mode, then all parameters end in e or E, as in h_{FE} and h_{oe}. The F and o stand for 'forward current gain' and 'output conductance', respectively: hence h_{FE} and h_{oe}. The capital letters stand for d.c. quantities and the lower case ones for a.c quantities. This standard is adopted by most engineers. The reciprocal of conductance is, of course, resistance Therefore, the collector current source output resistance is $1/h_{oe}$. Conductance is measured in a

unit which is the reciprocal of an ohm, called a 'siemen' (S). For instance, the reciprocal of $1\,M\Omega$ is a conductance of $10^{-6}S$ or $1\,\mu S$

Self-assessment questions

SAQ9.22 Increasing the collector resistance R_c of a CE circuit increases its voltage gain What happens to its output resistance r_{out}?

SAQ9.23 How can we increase A_v without affecting r_{out}?

SAQ9.24 A CE circuit has $R_c = 10\,k\Omega$ and a BJT with $h_{oe} = 5\,\mu S$. What effect has h_{oe} on the output resistance of the circuit? What would happen to r_{out} if h_{oe} were changed to $1\,\mu S$?

More about A_v and bypass capacitors
Recall that for a CE circuit

$$A_v \approx -R_c/(R_c + r_e)$$

Unfortunately, the values of R_c and R_e are determined by the Q point calculation! Look at the circuit in Figure 8 19 and the bias calculations in Example 8.5. Here, the Q point calculation has determined that $R_c = 7\,k\Omega$ and $R_e = 1\,k\Omega$. The approximate A_v of this circuit is therefore

$$A_v \approx -7\ k\Omega/1\ k\Omega$$

$$\approx -7$$

This leaves us with a rather inflexible A_v value, set at whatever the bias condition determines. Surely there must be a way of increasing the gain without altering the bias? There is, and this is where the bypass capacitor comes in. The circuit of Example 8.5 is shown in Figure 9 17 but, instead of $R_e = 1\,k\Omega$, we have shown it split into two series resistors R_{e1} and R_{e2} whose total series resistance is $1\,k\Omega$ We have also added a bypass capacitor C_e which has the following effect

1. At d c., C_e is an open circuit and $1\,k\Omega$ appears in the emitter circuit
2   At a c., the $900\,\Omega$ resistor is bypassed, leaving us with only $100\,\Omega$

At a c., therefore, the total emitter resistance is $(R_{e1} + r_e)$ and the gain equation becomes $A_v \approx -R_c/(R_{e1} + r_e)$, where $R_{e2} \approx 0\,\Omega$ as a result of the bypassing effect. Since the example quotes $I_{cq} = 1\,mA$, we can write the new gain equation as

$$A_v \approx -7\ k\Omega/[100\Omega + (26/1)\Omega]$$

$$\approx -56$$

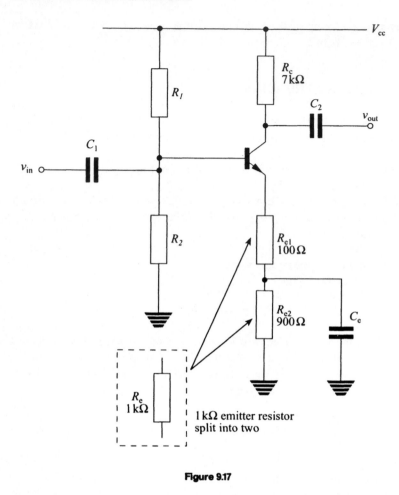

Figure 9.17

This is a large increase in voltage gain. Obviously, different values of R_{e1} and R_{e2} can be chosen to implement different gain values. The only stipulation is that their sum must be $1\,k\Omega$ (for this example).

Self-assessment questions

SAQ9.25 Choose emitter resistors to give a voltage gain of −20 in Example 8.5.

SAQ9.26 Choose an emitter bypass capacitor to bypass a 900 Ω resistor at a frequency range of 2–15 kHz.

Checklist

You should now be able to:
- draw and use the r_e BJT model;
- calculate r_e given the Q point;
- determine the input resistance of a BJT circuit with and without an external R_e;
- draw a.c. and d.c. equivalent circuits;
- understand how to treat capacitors and power supplies in equivalent circuits;
- use and calculate suitable values for coupling and bypass capacitors;
- estimate the approximate a.c. input and output resistances of CE and CC circuits;
- estimate the approximate voltage gain of a CE;
- modify a CE circuit to increase the voltage gain;
- design full CE and CC circuits to fit d.c. and a.c specifications.

Tutorial questions

1   Find the intrinsic emitter resistance of a BJT whose Q point is (18 V, 2 mA), (9 V, 100 μA) and (3 V, 1 A).
2   Determine the I_{cq} value of BJTs with r_e values of 360 Ω, 0.04 Ω, 1 kΩ and 150 Ω
3   What is the approximate Q point of a mid-point biased CC circuit with a 20 V power supply and r_e = 480 Ω, and a 6 V power supply and r_e = 5 Ω?
4   What is the input resistance of a BJT CC circuit with β = 50, Q point of (12 V, 2 mA) and R_e = 1 kΩ?
5   Find the input resistance and the collector voltage of a BJT CE circuit with β = 150, Q point of (8 V, 100 μA) and R_e = 10 kΩ.
6   If I_{cq} is increased, is input resistance also increased?
7   Why do we consider only the minimum frequency of a circuit when calculating coupling capacitor values?
8   Draw a simple series RC circuit. Now draw its a.c. and d.c. equivalent circuits.
9   If r_{in}' of a circuit is 50 kΩ, find a suitable value coupling capacitor in the frequency range 500–10 000 Hz. You may assume that the circuit providing the input signal is an ideal voltage source.
10 Find a suitable input coupling capacitor for an amplifier with a frequency response of 30 Hz to 12 kHz and r_{in}' = 90 kΩ, driven from a voltage source with r_{out} = 15 kΩ.
11 Determine a suitable output coupling capacitor for an amplifier with a frequency response of 1–12 kHz and r_{out} = 5 kΩ to drive a circuit with r_{in}' = 5 kΩ.

12   Find the approximate r_{out} value of a CC circuit with $R_e = 1\,k\Omega$ and $r_e = 100\,\Omega$.

13   Find the value of r_{out} of a mid-point biased CE circuit with $V_{cc} = 15\,V$, $I_{cq} = 1.5\,mA$ and $R_e = 2\,k\Omega$.

14   A CE circuit has $R_c = 25\,k\Omega$, $R_e = 2.5\,k\Omega$ and $I_{cq} = 0.5\,mA$. Its frequency response is to be 80 Hz to 8 kHz. Redesign the circuit without changing the bias values to increase the voltage gain to -20.

15   Fully design, using voltage divider biasing, a CE circuit with a.c. coupling and $V_{cc} = 18\,V$, $I_{cq} = 0.1\,mA$, $A_v = -40$, $100 \le \beta \le 500$ and a frequency response of 20 Hz to 10 kHz. (Find all R and C values required.)

16   Find the input and output resistances of the circuit you have designed in question 15.

17   Fully design, using voltage divider biasing, a CC circuit with a.c. coupling, a Q point of (6 V, 10 mA) and a frequency response of 30 Hz to 10 kHz. Assume the BJT has $50 \le \beta \le 200$. (Find all R and C values required.)

18   Imagine you have an oscilloscope to look at waveforms at different test points on the two circuits designed in questions 15 and 17. If we choose to drive the circuits with a 1 kHz sine wave with $V_{pk} = 50\,mV$ and zero d c. offset, draw diagrams of the waveforms seen at each node of the two circuits. (Label each node A, B, C, etc., to cross-reference the waveform diagram with the circuit diagram and remember to indicate the d.c. offset values clearly.)

Further reading

Bogart T.F. (1986) *Electronic devices and circuits*, sections 5 1, 5 3 and 5 6. Merrill, Columbus, Ohio

Ritchie G J (1987). *Transistor circuit techniques*, pp 43–7, 54–64 Van Nostrand Reinhold, Wokingham, England.

CHAPTER 10

MOSFETs

Aims and objectives

The aim of this chapter is to introduce the MOSFET as an alternative device to the BJT. The chapter briefly covers MOSFET operation with reference to the depletion and enhancement modes. The aim is to give a descriptive account of the device to enable the reader to make comparisons with the BJT and to see how it is used in digital electronics. It is important in this chapter to introduce the reader to some of the terms commonly used in circuits using MOSFET technology. Detailed MOSFET circuit configurations and analyses are not covered.

After reading this chapter you will be more conversant with:

- *MOSFET operation*
- *enhancement mode and depletion mode devices*
- *V_{gs}, V_{ds} and I_d, I_{dss}, V_{TH}, $V_{gs(off)}$ and R_{ds}*
- *NMOS and PMOS circuits*
- *active and passive loads*
- *CMOS circuits*
- *handling MOSFETs*
- *how the MOSFET compares with the BJT*

The metal oxide semiconductor field effect transistor, to give the MOSFET its full title, is one of many members of the field effect transistor (FET) family. The FET family is constantly being developed using new materials and smaller device areas. The reason for introducing it here is to whet your appetite for further studies of MOSFET circuits used in digital electronics. Its use is by no means confined to digital circuits, but seeing it in use in a logic circuit will help your understanding of its operation. We will look at the two MOSFETs used in circuits· the E-MOSFET and the D-MOSFET

10.1 The enhancement mode MOSFET

The enhancement mode MOSFET (E-MOSFET) is usually a four-terminal device

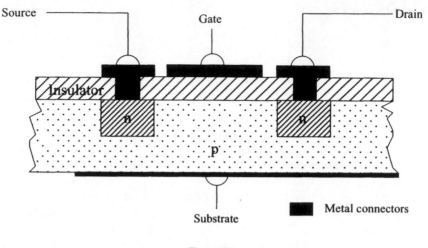

Figure 10.1

with a gate (g), source (s), drain (d) and substrate connection. Its layout is shown in Figure 10.1. Usually the substrate connection is internally connected to the source, making it a three-terminal device (just like the BJT). One of the big differences between this and the BJT is that one of its connections, the gate, is not connected directly to the semiconductor at all: it is connected to a metal plate on one side of an insulating layer and is therefore insulated from the rest of the device. For this reason the device is often called an 'insulated gate FET' (IGFET).

The basic device layout consists of a large piece of semiconductor material called the substrate (shown as p-type) into which two pieces of the opposite type are diffused (shown as n-type), one of which is a drain, the other a source. The device is then covered with an insulating layer and finally a metal gate connection is added to the top of the insulator, along with metal connectors through the insulator to connect to the diffused drain and source regions.

A quick look at the E-MOSFET symbol shown in Figure 10.2 tells us that there are two versions available: an n channel and p channel. Before looking at how the

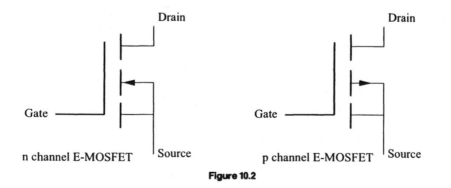

Figure 10.2

device works, note that the symbol shows the internal connection between substrate and source The n channel has a p-type substrate and n-type drain and source regions, whereas the p channel swaps the types around.

10.1.1 How the E-MOSFET works

So how can a device work when one of its connectors is insulated from the rest? The key to its operation is in the name, 'the field effect' Device operation comes about from the electric field produced when a voltage is applied to the gate. Let us consider the n channel and see what happens. Look at Figure 10.3.

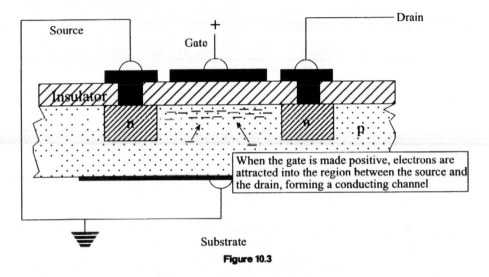

When the gate is made positive, electrons are attracted into the region between the source and the drain, forming a conducting channel

Substrate

Figure 10.3

If a voltage is applied to the gate which is positive with respect to the substrate, an electric field is produced. This field attracts any electrons in the substrate up to the region underneath the insulator. The collection of electrons forms a channel between the n-type drain and source regions, allowing a current to flow from the drain to the source or vice versa. This collection of electrons in the channel effectively changes the material (in this region) from p-type to n-type and is called an 'inverted channel'.

10.1.2 Enhancement

Because the action of the electric field improves the conduction of the channel, it is called an enhancement mode device (enhance merely means increase in value). Once a gate voltage is reached to produce an inverted channel, increasing the gate voltage further causes more drain current to flow. The amount of gate voltage required to cause inversion is called the 'threshold voltage' (V_{TH}) and for an n channel device is typically a few volts (1–5 V)

Self-assessment question

SAQ10.1 An E-MOSFET has $V_{TH} = +2\,V$. If we apply $V_{gs} = +1\,V$, will the channel be inverted?

10.1.3 Gate current

Because the gate is insulated, the device operation requires no d c. current to flow in the input to the gate. This makes the device a valuable addition to any electronics design as it has an enormous input resistance $R_{in} > 10^{12}\Omega$.

10.1.4 Controlling drain current

The input gate voltage controls the amount of output drain current flow and is therefore useful as a voltage-to-current (or transconductance) amplifier.

10.1.5 I_d versus V_{gs}

In Figure 10.4 we can see a circuit showing a variable gate source voltage (V_{gs}) controlling a drain current (I_d) and a plot showing the increase in I_d with V_{gs}. The increase in I_d with V_{gs} is not linear but parabolic. The mathematics for this relationship is dependent on the actual device used. If a device is tested and a measurement made of drain current ($I_{d(on)}$) at a specific gate source voltage ($V_{gs(on)}$), as shown in Figure 10.4, well above V_{TH}, the expression for I_d at any V_{gs} becomes

$$I_d = I_{d(on)} \times [(V_{gs} - V_{TH})^2/(V_{gs(on)} - V_{TH})^2]$$

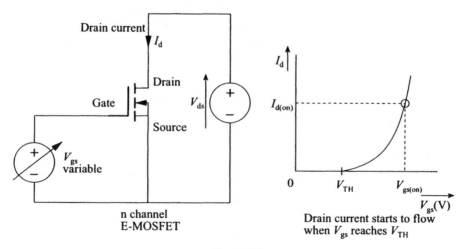

Figure 10.4

Although this expression is a little messy, finding I_d for any V_{gs} value is just a matter of putting numbers into the equation Let us try an example.

EXAMPLE 10.1
Find the drain current at $V_{gs} = +5\,V$ for an E-MOSFET with $V_{TH} = +1\,V$, $I_{d(on)} = 5\,mA$ and $V_{gs(on)} = 10\,V$.

SOLUTION
$$I_d = 5\ mA \times [(5\ V - 1\ V)^2/(10\ V - 1\ V)^2]$$

$$\approx 0.99\ mA$$

10.1.6 I_d versus V_{ds}

Now we know that a specific gate voltage V_{gs} will produce a specific I_d value, we can fix the V_{gs} value and see how varying V_{ds} will affect the drain current. The circuit in Figure 10.5 shows this V_{gs} is kept constant and V_{ds} is increased from zero As V_{ds} increases, I_d increases linearly until $I_{d(on)}$ is reached Further increases in V_{ds} have no effect on I_d and its value remains constant. In other words, a constant V_{gs} gives a constant I_d, provided V_{ds} does not drop too low. Varying V_{ds} produces a plot, shown in Figure 10.5, with two distinct parts.

1. a steady slope at low V_{ds}, called the 'ohmic' or 'linear' region;
2. a horizontal, constant I_d at larger V_{ds}, called the 'saturation' or 'constant current' region.

These two regions define two modes of device operation.

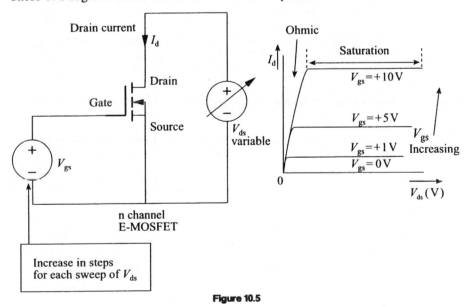

Figure 10.5

Self-assessment questions

SAQ10.2 Find the drain current at $V_{gs} = +8$ V for an E-MOSFET with $V_{TH} = +2$ V, $I_{d(on)} = 1$ mA and $V_{gs(on)} = 5$ V.

SAQ10.3 What size is the input current to the gate in SAQ10.2?

In ohmic or linear mode the E-MOSFET can be used as a resistance since an increase in V_{ds} will cause a linear increase in I_d

In saturation or constant current mode the device will conduct a steady current, the size of which is dependent on the size of V_{gs}.

10.1.7 At what V_{ds} value does the ohmic region become the saturation region?

The breakpoint between the two parts of the graph is when V_{ds} is just equal to the difference between V_{gs} and V_{TH}, i.e. the amount that V_{gs} is above V_{TH} Therefore, the graph changes from linear to saturation at

$$V_{ds} = V_{gs} - V_{TH}$$

If $V_{ds} < (V_{gs} - V_{TH})$, the E-MOSFET is in linear or ohmic mode and if $V_{ds} > (V_{gs} - V_{TH})$, it is in saturation or current mode.

Self-assessment questions

SAQ10.4 An E-MOSFET has $V_{TH} = +2$ V, $V_{ds} = +3$ V and $V_{gs} = +4$ V. Which mode of operation is it in?

SAQ10.5 An E-MOSFET has $V_{TH} = +3$ V and $V_{ds} = +10$ V. What size of V_{gs} is necessary to keep the device in ohmic mode?

SAQ10.6 Using the I_d versus V_{ds} plot, what can you say about the resistance between the drain and the source in the saturation region? Does it change with V_{ds}?

10.1.8 The drain to source resistance

When a positive gate voltage is applied to the E-MOSFET, the conducting channel causes the resistance between the drain and the source (R_{ds}) to drop from a very high value (about 10^6 Ω) to a very low value (about 10 Ω). This R_{ds} value can usually be found from the data sheet of the actual device being used. R_{ds} is at its lowest value when the E-MOSFET is operating in its ohmic region

10.1.9 The MOSFET switch

The ability of the E-MOSFET to switch R_{ds} from high to low by the application of an input voltage to an almost infinite resistance gate makes it the most useful and commonly used switch ever produced. Computer circuits at present use MOSFET technology where millions of switches are in action in each integrated circuit every microsecond.

10.1.10 The two-terminal E-MOSFET resistor

There are times when it is useful to make the E-MOSFET behave as a resistor. However, although the E-MOSFET can be used in its ohmic mode, for it to be of any use we must prevent it entering its saturation mode. This might be difficult if V_{ds} is ever more than a few volts. To get round this, a clever technique is used. If the gate and the drain terminals are connected together, the I_d versus V_{ds} curve takes on the same shape as the I_d versus V_{gs} curve (see Figure 10.6) The relationship relies on the fact that an increase in V_{gs} results in an increase in I_d. Although we can see that this is not actually a perfect resistor, it performs well enough to warrant its extensive use in digital circuits. See Section 10.3 on MOSFET resistors.

Self-assessment question

SAQ10.7 Using the plot in Figure 10.6, say what happens to the resistance as V_{ds} increases

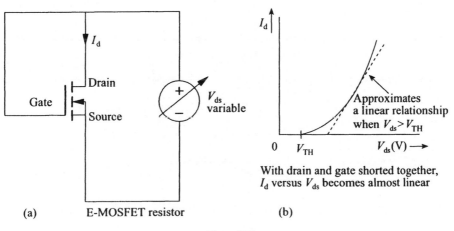

(a)   E-MOSFET resistor   (b)

With drain and gate shorted together,
I_d versus V_{ds} becomes almost linear

Figure 10.6

10.1.11 Summary of E-MOSFET action

When $V_{gs} < V_{TH}$, I_d is zero and R_{ds} is very large. When $V_{gs} > V_{TH}$, drain current flows and is given by

$$I_d = I_{d(on)} \times [(V_{gs} - V_{TH})^2/(V_{gs(on)} - V_{TH})^2]$$

R_{ds} drops to a low value.

When I_d flows, the E-MOSFET is in one of two modes

1. If $V_{ds} < (V_{gs} - V_{TH})$, it is in linear or ohmic mode.
2. If $V_{ds} > (V_{gs} - V_{TH})$, it is in saturation or current mode.

10.2 The depletion mode MOSFET

The depletion mode MOSFET (D-MOSFET) differs only very slightly in structure and operation from the E-MOSFET and is often used in conjunction with an E-MOSFET in a circuit. Look at Figures 10.7 and 10.8 to see the differences. Much of the description of the E-MOSFET refers also to the D-MOSFET.

The basic device layout consists of a substrate (shown as p-type) into which n-type drain and source regions are diffused. The D-MOSFET has, in addition, a lightly doped channel of n-type material linking the source and drain regions. The device is then covered with an insulating layer and finally a metal gate connection is added to the top of the insulator, along with metal connectors through the insulator to connect it to the diffused drain and source regions.

The D-MOSFET symbols in Figure 10.8 shows the n and p channel versions of the device. The symbol shows the internal connection between the substrate and the

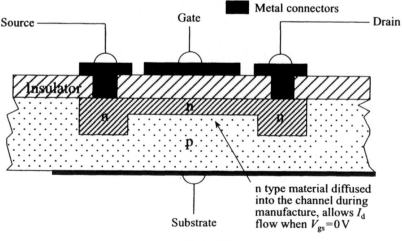

Figure 10.7

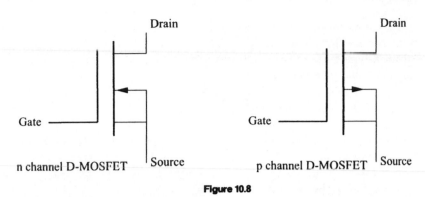

Figure 10.8

source, and the solid vertical line suggests the extra conducting channel diffused into the device Note the difference between these and the E-MOSFET symbols.

10.2.1 How the D-MOSFET device works

Let us consider the n channel and see what happens. Look at Figure 10.9.

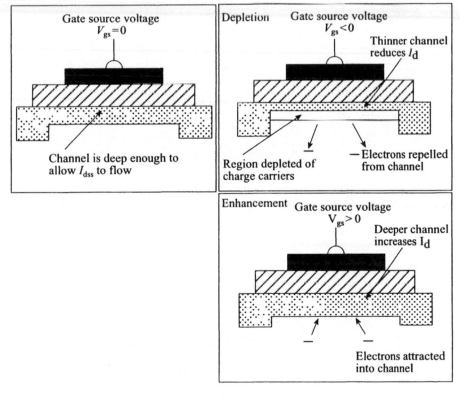

Figure 10.9

When $V_{gs} = 0$, we have a situation where a positive V_{gs} voltage is not required to cause drain current flow, since at $V_{gs} = 0$ there is already a conducting channel caused by the n-type diffused layer and a drain current called I_{dss} flows. What, then, is the purpose of the gate voltage?

When $V_{gs} > 0$, V_{gs} is positive, and the effect is identical to that of the E-MOSFET, i.e. increasing V_{gs} increases I_d.

When $V_{gs} < 0$, V_{gs} is negative and the drain current is reduced. If V_{gs} is made more negative, I_d continues to reduce until we reach a point where $I_d = 0$ The gate voltage required to do this is called $V_{gs(off)}$. The main difference, therefore, between this and the E-MOSFET is that we can increase or decrease I_d by applying either a positive or a negative gate voltage respectively.

10.2.2 Enhancement

The device used in its enhancement mode is identical to the description already given for the E-MOSFET Enhancement occurs whenever $V_{gs} > 0\,\text{V}$.

10.2.3 Depletion

When a negative V_{gs} is applied to the device, the electrons in the n-type channel are repelled away into the substrate and holes are attracted into the channel. This depletes the channel of electron charge and reduces the depth of the channel between the drain and the source. This in turn decreases the size of drain current flow. Larger negative values of V_{gs} deplete the channel of charge and prevent drain current flow altogether.

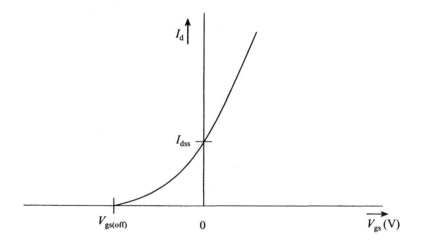

Figure 10.10

10.2.4 I_d versus V_{gs}

The plot of I_d versus V_{gs} is the same shape as the E MOSFET, but the whole graph is shifted to the left. Look at Figure 10.10 Now I_d starts rising at $V_{gs(off)}$, which is negative. We can see from the plot that when $V_{gs} = 0$ the drain current is I_{dss}.

The expression for I_d at any V_{gs} is

$$I_d = I_{dss} \times [1 - (V_{gs}/V_{gs(off)})]^2$$

where I_{dss} is found from the data sheet for the actual device in use

EXAMPLE 10.2

Find the drain current at $V_{gs} = +3\,V$ for a D-MOSFET with $V_{gs(off)} = -2\,V$ and $I_{dss} = 2\,mA$

SOLUTION

Using $I_d = I_{dss} \times [1 - (V_{gs}/V_{gs(off)})]^2$, we have

$$I_d = 2\text{ mA} \times [1 - (+3/-2)]^2$$
$$\approx 12.5\text{ mA}$$

10.2.5 I_d versus V_{ds}

A plot of I_d versus V_{ds} (Figure 10 11) produces the same curves as are seen with the E-MOSFET. Note, however, the difference in the V_{gs} values, which can now be negative or positive above $V_{gs(off)}$ As with the E-MOSFET, the device can be operated in two distinct regions. ohmic or saturation (constant current)

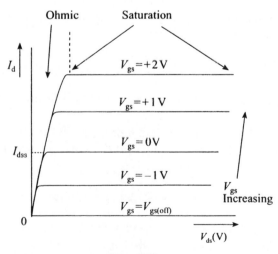

Figure 10.11

10.2.6 Summary of D-MOSFET action

D-MOSFETs can be used with positive and negative V_{gs} values. Provided that $V_{gs} > V_{gs(off)}$, drain current will flow. The drain current flow is given by

$$I_d = I_{dss} \times [1 - (V_{gs}/V_{gs(off)})]^2$$

When $V_{gs} = 0$, a data sheet specified drain current flows $I_d = I_{dss}$.
When $V_{gs} > 0$

$$I_d > I_{dss}$$

The device is in enhancement mode and behaves like an E-MOSFET
    When $V_{gs} < 0$

$$I_d < I_{dss}$$

The device is in depletion mode.
    When $V_{gs} = V_{gs(off)}$

$$I_d = 0$$

Self-assessment questions

SAQ10.8 Find the drain current at $V_{gs} = +6$ V for a D-MOSFET with $V_{gs(off)} = -4$ V and $I_{dss} = 5$ mA.

SAQ10.9 If $I_d = I_{dss}$ for a D-MOSFET, what value is V_{gs}?

SAQ10.10 In SAQ10.8, what value of V_{gs} would switch the MOSFET off?

SAQ10.11 Without performing any calculations, determine whether $V_{gs} = -3$ V for the device in SAQ10.8 would result in an I_d value of less than, or more than, I_{dss}.

SAQ10.12 A MOSFET has $V_{TH} = +4$ V Is it a depletion or an enhancement device?

SAQ10.13 A MOSFET has $V_{gs(off)} = -3$ V. If $V_{gs} = +2$ V, is it in enhancement or depletion mode?

10.3 MOSFETs as resistors in integrated circuits (chips)

Because of the MOSFET's popularity in digital circuits as a switching device, the need for smaller device areas and higher device counts per chip meant that their role in playing the part of a humble resistor was investigated.

Why use a MOSFET as a resistor when we could use a resistor instead?

The problem with producing a simple resistor in n- or p-type material is that a large resistance value requires a large area of material. A single integrated resistor could easily take up the same space as many MOSFET devices. The solution, therefore, is to use a MOSFET (which takes up little space) as a resistor

In digital circuitry, the role of the resistor is usually to act as a load in the drain circuit. The circuit used most often is the inverter, which is shown in Figure 10.12. Figure 10.12(a) shows the traditional inverter with a simple passive resistor as the load. Figure 10 12(b) and 10.12(c) shows the passive load replaced by depletion and enhancement loads respectively We will assume that V_{in} can take only two values: 0 V or +5 V. Let us take a brief look at the circuit action, remembering that V_{TH} is positive for all the n channel E-MOSFETs. We will assume that $V_{TH} = +1$ V.

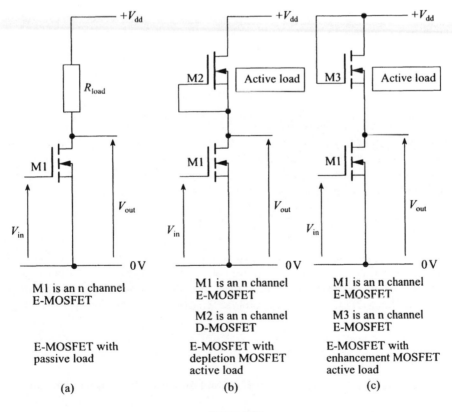

M1 is an n channel E-MOSFET	M1 is an n channel E-MOSFET	M1 is an n channel E-MOSFET
	M2 is an n channel D-MOSFET	M3 is an n channel E-MOSFET
E-MOSFET with passive load	E-MOSFET with depletion MOSFET active load	E-MOSFET with enhancement MOSFET active load
(a)	(b)	(c)

Figure 10.12

For the circuit in Figure 10.12(a), when $V_{in} = 0$, M1 is off and $I_d = 0$. Since M1 and R_{load} are in series, no voltage is dropped across R_{load} and

$$V_{out} = V_{dd}$$

When V_{in} is positive, M1 conducts, I_d flows and the R_{load} value has been chosen to drop almost all of V_{dd}, i.e.

$$V_{out} \approx 0 \text{ V}$$

For the circuit in Figure 10.12(b), M2 (D-MOSFET) has $V_{gs} = 0$ and therefore will operate somewhere on the $V_{gs} = 0$ plot in Figure 10.11. When $V_{in} = 0$, M1 is off and therefore $I_d = 0$. Since M1 and M2 are in series, M2 also has $I_d = 0$. With reference to Figure 10.11, the only place on the $V_{gs} = 0$ plot where $I_d = 0$ is when $V_{ds} = 0$. Therefore, there is no voltage drop across the D-MOSFET and

$$V_{out} = V_{dd}$$

When V_{in} is positive, M1 conducts, I_d flows and M2 carries the same current. Again, with reference to Figure 10.11, the only place on the $V_{gs} = 0$ plot where $I_d > 0$ is when $V_{ds} > 0$. M2 limits the drain current flow to I_{dss} and drops all of V_{dd}. Therefore

$$V_{out} \approx 0 \text{ V}$$

For the circuit in Figure 10.12(c), M3 (E-MOSFET) has its drain and gate connected together and will therefore operate as a resistor, as shown in the I_d versus V_{ds} curve in Figure 10 6. When $V_{in} = 0$, M1 is off and therefore $I_d = 0$. Since M1 and M3 are in series, M3 also has $I_d = 0$. With reference to Figure 10.6, the only place on the curve where $I_d = 0$ is when $V_{ds} = 0$ or at a low voltage. Therefore, there is little or no voltage drop across the E-MOSFET resistor and

$$V_{out} \approx V_{dd}$$

When V_{in} is positive, M1 conducts, I_d flows and M3 carries the same current Again with reference to Figure 10.6, the only place on the curve where $I_{d1} > 0$ is when $V_{ds} > 0$. The M3 E-MOSFET resistor is designed to drop almost all of V_{dd}. Therefore

$$V_{out} \approx 0 \text{ V}$$

In the circuits shown in Figure 10 12(a) and 10.12(b), the passive load resistor has been replaced by a MOSFET device which is behaving as a resistor When devices are used in this way they are called 'active loads'.

10.4 NMOS and PMOS circuits

The active load inverter circuits shown here all use exclusively n channel MOSFETs only Circuits of this type are generally referred to as NMOS circuits.

Although we have not considered any p channel devices, all the descriptions and circuits shown so far could be altered for p channel devices. The difference is that a p channel device requires a negative V_{gs} to switch it on and current flows from the source to the drain. For this reason, all p channel circuits have power supplies of opposite polarity from the n channel. Circuits made up exclusively of p channel devices are called PMOS circuits.

10.5 CMOS circuits

The NMOS and PMOS inverters have been successfully used in many digital circuits, although the NMOS device is more commonly used because it switches faster than the PMOS. However, circuits which use a complementary mixture of NMOS and PMOS devices have become popular in battery powered appliances because of their extremely low power requirements.

These circuits are called CMOS from the description 'complementary MOSFET' Complementary means 'completing or making up a whole' and in this context it means that when the NMOS is off, the PMOS is on, and vice versa. So what does the CMOS inverter look like and why does it require so little power? Let us see. Figure 10 13 shows the circuit. The CMOS inverter works like this.

Assume V_{in} can be either 0 V or 5 V. When $V_{in} = 0$ V, M1 is off because $V_{gs1} = 0$ V and therefore $I_d = 0$. M2 is on because $V_{gs2} = -5$ V, but what size is I_d? Since M1 and M2 are in series, $I_d = 0$ through both With reference to the p channel I_d/V_{ds} curve shown in Figure 10.13(b), the only point on the $V_{gs2} = -5$ V curve where $I_{d2} = 0$ is when $V_{ds2} = 0$ V. Therefore, M2 is on and $V_{ds2} = 0$, M1 is off and $V_{ds1} = V_{dd}$ Therefore

$$V_{out} = V_{dd}$$

When $V_{in} = +5$ V, M2 is off because $V_{gs2} = 0$ V and therefore $I_d = 0$ M1 is on because $V_{gs1} = +5$ V, but what size is I_d? Since M1 and M2 are in series, $I_d = 0$ through both. Therefore, with reference to the n channel I_d/V_{ds} curve shown in Figure 10 13(c), the only point on the $V_{gs1} = +5$ V curve where $I_{d1} = 0$ is when $V_{ds1} = 0$ V Therefore, M1 is on and $V_{ds1} = 0$, M2 is off and $V_{ds2} = V_{dd}$. Therefore

$$V_{out} = 0$$

This explains the inverting action of the CMOS circuit.

Note the important part, which is that although M1 and M2 are on and off alternately, I_d is zero in each case! We would therefore make the logical deduction

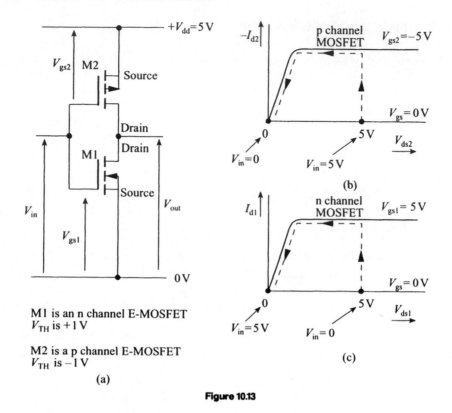

Figure 10.13

that if the circuit draws no current, the power supply providing V_{dd} will last for ever! This is not quite true (as you might have guessed).

Although it is true that no current is drawn after each switching action has occurred, the circuit draws some current during its change from one state to another. An ammeter in the power supply line to V_{dd} would register a little blip of current while V_{out} changed levels Otherwise, it is an extremely low power circuit and is used because of it.

10.6 PMOS, NMOS and CMOS handling precautions

The gate of a MOSFET device, as we have already discussed, is connected to an insulated gate and requires no input current. Although this is good as far as electronic design goes, it makes the device very sensitive to static electricity. Static electricity, by virtue of its name, is a voltage build-up without any current flow (static, unmoving charge). This means that static voltages of many thousands of volts can build up on the MOSFET gate terminal, with nowhere for it to flow away as current. If the voltage gets too high, the thin insulating oxide between the gate and the substrate can break down to destroy the device.

Therefore, if using MOSFET circuits, always take the following precautions:

1. Make sure a conductive foam material joins all the terminals together until the circuit is plugged in and powered up.
2. Earth yourself with an earthing strap (connected to a known earth point) before handling the circuit.
3. Avoid touching the circuit terminals at any time.
4 Never unplug a MOSFET circuit and leave it unconnected on a bench or worktop.
5. Avoid direct soldering of a MOSFET device or circuit Try to use a socket which can be soldered before inserting the circuit

10.7 Comparing a MOSFET with a BJT

Terminals
BJTs have three terminals: collector, base and emitter.

MOSFETs have three corresponding terminals: drain, gate and source. A fourth terminal, the substrate, is often connected internally to the source.

Input voltage
A MOSFET usually requires a few volts of input V_{gs} to switch it on.

A BJT only requires $V_{be} \approx 0.6\,\text{V}$ to switch it on.

Input resistance
BJT circuits *can* be designed with high input resistances.

MOSFETs easily have the highest input resistance of any circuits

Input current
BJTs require a few microamps or milliamps input current to operate.

MOSFETs require no input current to operate

Voltage gain
BJT circuits can be designed with medium to large voltage gains since a small increase in V_{be} causes a large increase in collector current

MOSFET circuits do not have large voltage gains since a large variation in V_{gs} is required to obtain changes in drain current.

Current gain
BJTs have medium current gain equal to h_{FE}

MOSFETs have almost infinite current gain since no input current is required.

Current flow through the device
A BJT has no current flow until an input voltage is applied.

An E-MOSFET has no current flow until an input voltage is applied. A D-MOSFET has a drain current flow called I_{dss} with zero V_{gs} input.

Output resistance
BJT circuits can be designed to have very low output resistances.

MOSFET circuits typically do not exhibit very low output resistance values.

Capacitance
BJT input capacitance is very low and is usually not a big problem, except at high frequencies.

MOSFETs have large input capacitance values because of the gate/insulator/substrate capacitor-type structure. This is often a problem requiring consideration, even at medium frequencies, as the gate capacitor draws a short-lived charging current before switching

Speed of operation
BJT circuits can switch at high speed.

MOSFET switching speed is reduced by the capacitance effects and higher output resistances.

Handling
BJTs need no special handling considerations.

MOSFETs can be damaged by static and require special handling.

Self-assessment questions

SAQ10.14 When does I_d flow in an NMOS inverter?

SAQ10.15 When does I_d flow in a CMOS inverter?

SAQ10.16 Can a MOSFET be switched on and yet carry no drain current?

SAQ10.17 Would we generally select a BJT or a MOSFET for circuits with the following characteristics?
1. very high R_{in},
2. very low R_{out},
3. very high A_v;
4. very high A_i;
5. a very fast switching speed,
6. a very low input capacitance at low frequencies;
7. not being static-sensitive;
8. battery-operated switching circuits.

SAQ10.18 Why are active loads used?

SAQ10.19 A p channel E MOSFET has $V_{TH} = -3\,V$. If $V_{gs} - 0$, is the device on or off?

SAQ10.20 Can you explain why an op-amp might have a MOSFET input stage followed by subsequent BJT stages?

Checklist

You should now know the following
- what FET and MOSFET stand for;
- how to draw the D-MOSFET and E-MOSFET n and p channel symbols;
- what the difference between a D-MOSFET and an E-MOSFET is,
- what enhancement, depletion and inversion mean,
- why the MOSFET is sometimes called IGFET,
- what a typical V_{TH} value would be for a p or an n channel MOSFET;
- how to sketch the I_d/V_{gs} and I_d/V_{ds} plots for all the MOSFET types;
- the difference between V_{TH} and $V_{gs(off)}$,
- which plot indicates the MOSFET's ohmic region;
- which device has a drain current of I_{dss},
- how to turn off an E-MOSFET;
- how to turn off a D-MOSFET;
- how to connect a MOSFET to act as a two-terminal resistor,
- what an active load is,
- why the names NMOS, PMOS and CMOS are used,
- why a CMOS inverter draws almost no current from its supply,
- how to compare BJT and MOSFET circuit characteristics,
- why MOSFET circuits should be handled with care.

Tutorial questions

1 An E-MOSFET has $V_{TH} = +4\,V$ If $V_{gs} = +5\,V$ is applied, will there be a drain current flow?
2 Find I_d for an E-MOSFET with $V_{TH} = +4\,V$, $I_{ds(on)} = 20\,mA$, $V_{gs(on)} = 15\,V$ and $V_{gs} = 10\,V$
3 An E-MOSFET has $V_{TH} = +2\,V$. Is it an n or a p channel device?
4 An E-MOSFET has $I_d = 3\,mA$, $V_{TH} = +2\,V$, $V_{gs(on)} = 5\,V$ and $I_{d(on)} = 5\,mA$ What is the applied V_{gs} value?
5 What is the size of I_d in question 4 if V_{gs} is raised to $+10\,V$?
6 An E-MOSFET has $V_{TH} = +8\,V$, $V_{gs} = +10\,V$ and $V_{ds} = +9\,V$. Is it in linear or saturation mode? What range of V_{ds} values would keep it in linear mode?

7   If an n channel E-MOSFET is in saturation mode, will an increase in V_{ds} result in an increase or a decrease in I_d?

8   If an n channel D-MOSFET has $V_{gs} > 0$, is $I_d > I_{dss}$?

9   Find I_d at $V_{gs} = +20$ V for an n channel D-MOSFET with $V_{gs(off)} = -5$ V and $I_{dss} = 4$ mA.

10   Find I_d for the device in question 9 if V_{gs} drops to $-4\,5$ V.

11   Can we *always* assume a drain current is flowing for an n channel MOSFET when $V_{gs} > V_{TH}$? Can you explain your answer?

12   Compare the advantages and disadvantages of BJT and CMOS inverters

Further reading

Bogart T F (1986) *Electronic devices and circuits*, section 7.5 Merrill, Columbus, Ohio
Millman J & Halkias C (1988). *Integrated electronics*, pp 322–44 McGraw-Hill, Singapore

Answers to self-assessment questions

Chapter 1

SAQ1.1
1. Should be '.   taken across the resistor R . . '
2. Should be '. . . the current through the bulb   .'
3. Should be '.   of 5 5 V across the resistance .   '

SAQ1.2
1. True
2. False
3. False
4. False.
5. True
6. False.

SAQ1.3
In series R_{total} − 90 kΩ, in parallel $R_{total} = 6$ 3 kΩ (6 269).

SAQ1.4
$1/R_{total} = 1/R_1 + 1/R_2 = (R_2 + R_1)/R_1 \times R_2$. Therefore, $R_{total} = R_1 \times R_2/(R_2 + R_1)$

SAQ1.5
Current enters from the left-hand side, therefore, the left-hand side of the resistor is positive

SAQ1.6
kΩ

SAQ1.7
Circuit current $I = 2/(10\ \text{k}\Omega + 50\ \text{k}\Omega) = 0$ 033 mA $(= 33\ \mu\text{A})$ Therefore·
1. 33 μA
2. 33 μA
3. $V_{10\ \text{k}\Omega} = I \times R = 0$ 033 mA $\times 10\ \text{k}\Omega = 0.33$ V
4. $V_{RL} = 2 - 0$ 33 = 1 67 V

SAQ1.8

1. $I_1 = I_{in} \times R_2/(R_1 + R_2) = 15 \text{ mA} \times 25 \text{ k}\Omega/35 \text{ k}\Omega = 107 \text{ mA}$. Therefore, $V_{R1} = I_1 \times R_1 = 1.07 \text{ mA} \times 10 \text{ k}\Omega = 10.7 \text{ V}$.

2. $V_{R2} = V_{R1} = 107 \text{ V}$

3. $I_1 = 107 \text{ mA}$, $I_2 = 15 \text{ mA} - 107 \text{ mA} = 043 \text{ mA}$ There are many other ways of answering this question.

SAQ1.9

1. $V_{out} = 15 \times 12 \text{ k}\Omega/(12 \text{ k}\Omega + 24 \text{ k}\Omega) = 5 \text{ V}$

Similarly·

2. 2 V

3. 5 V

4. 2 V.

SAQ1.10

1. Yes

2. Yes ($V_{R3} = 12 \times 15 \text{ k}\Omega/36 \text{ k}\Omega = 5 \text{ V}$)

3. Yes

SAQ1.11

The lost 2.5 V has been dropped across the internal resistance of the battery No, it would not be possible to read the 2 5 V because the internal resistance is, by definition, internal to the battery itself, made up of a combination of chemical and electrode resistances!

SAQ1.12

1. Power $P = V \times I$ Therefore, for the lamp where $V = 6$ V and $I = 05$ A, we have $P = (6 \text{ V}) \times (05 \text{ A}) = 3 \text{ W}$.

2. First, forget about the time (2 hours), which is added only to make the problem seem realistic If the terminal voltage has dropped to 4 V, we must therefore have 2 V dropped across the internal battery resistance, $V_{int} = 2 \text{ V}$ If $R_{lamp} = 12 \, \Omega$ and $V_{lamp} = 4$ V, Ohm's law gives $I_{lamp} = 4 \text{ V}/12 \, \Omega$. This current also flows through the battery resistance (a series current). Therefore, $I_{lamp} = I_{battery} = (4/12) \text{ A}$. From Ohm's law, $R_{int} = V_{int} / I_{battery} = 2 \text{ V}/(4/12) \text{ A} = 6 \, \Omega$

3. Assuming a new battery to be a good voltage source, $R_{int} \approx 0 \, \Omega$

4. $0 \, \Omega$

SAQ1.13

If the microphone is a poor voltage source, its R_{int} is high (or very high) As soon as load current flows, the internal voltage drop is also high, causing a greatly reduced signal voltage to the load. It is an example of poor voltage transfer.

SAQ1.14

Referring to Figure 1.26, we are given $I_T = 100 \text{ mA}$, $R_{int} = 10 \text{ k}\Omega$, $R_L = 150 \, \Omega$ Find I Using the current divider equation: $I = I_T \times R_{int}/(R_{int} + R_L) = 100 \text{ mA} \times 10\,000 \, \Omega/(10\,000 \, \Omega + 150 \, \Omega) = 98\ 5 \text{ mA}$ We would assume that the parameters describing the current source are fixed, leaving us only one option: to obtain a larger load current, we would try to reduce R_L

SAQ1.15
$\infty \ \Omega$

SAQ1.16
The larger internal resistance works best as a current source; therefore, choose
$R_{int} = 110 \ k\Omega$

Chapter 2

SAQ2.1
We assume that the voltage and current sources used in the circuit are ideal and
therefore contain their appropriate ideal internal resistances. $R_{int} = 0 \ \Omega$ for a
voltage source, which is equivalent to a short circuit, and $R_{int} = \infty \ \Omega$ for a current
source, which is equivalent to an open circuit.

SAQ2.2
We have already found the TEC for Figure 2 9; $V_{th} = 12 \ 8 \ V$ and $R_{th} = 9 \ k\Omega$.
Therefore, with a new load of 1 kΩ, we would have a new load voltage V_L, given by

$$V_L = 12.8 \ V \times (1 \ k\Omega)/(1 \ k\Omega + 9 \ k\Omega) = 1.28 \ V$$

SAQ2.3
Having an R_{th} value of 9 kΩ is the same as saying its internal resistance is 9 kΩ It
would not, therefore, be considered a good voltage source. Voltage transfer would
only be suitable for loads of 90 kΩ or more!

SAQ2.4
The TEC determined for this circuit has $V_{th} = 15 \ V$ and $R_{th} = 11.25 \ k\Omega$. An 11.25 kΩ
load added to this circuit would merely halve the load voltage to 15/2 V, i.e. $V_L =$
7 5 V. This should be reasonably obvious without performing a voltage divider
calculation.

SAQ2.5
$V_L = 15 \times 50 \ k\Omega/(50 \ k\Omega + 11.25 \ k\Omega) \approx 12 \ 25 \ V$

SAQ2.6
Having an R_{th} value of 11.25 kΩ, it would not be considered a good voltage source.
Voltage transfer would only be suitable for loads of 113 kΩ or more!

SAQ2.7
The TEC can be found from your diagram as follows

$$R_{th} = 22 \ k\Omega \parallel 48 \ k\Omega = 22 \times 48/(22 + 48) = 15.09 \ k\Omega$$

$$V_{th} = 15 \times 48 \ k\Omega/(48 \ k\Omega + 22 \ k\Omega) = 10.29 \ V$$

1   $V_L = 10 \ 29 \times (3 \ k\Omega)/(3 \ k\Omega + 15.09 \ k\Omega) \approx 1 71 V$
2   $V_L = 10 \ 29 \times (10 \ k\Omega)/(10 \ k\Omega + 15.09 \ k\Omega) \approx 4.1 V$

3. $V_L = 10\ 29 \times (48\ k\Omega)/(48\ k\Omega + 15\ 09\ k\Omega) \approx 7.83\ V$
4. $V_L = 10.29 \times (200\ k\Omega)/(200\ k\Omega + 15\ 09\ k\Omega) \approx 9\ 57\ V.$

SAQ2.8

The I_{Norton} of Figure 2.2 can be found by short circuiting the 30 kΩ resistor and finding the resulting current through 18 kΩ: $I_{Norton} = 24\ V/18\ k\Omega = 1\ 33$ mA and $R_{Norton} = 18\ k\Omega \parallel 30\ k\Omega = 11.25\ k\Omega$

1. $I_L \approx 1\ 33/2 \approx 0\ 67\ mA; V_L \approx (0\ 67\ mA) \times (11.25\ k\Omega) \approx 7\ 5\ V$
2. $I_L \approx (1\ 33) \times (11\ 25\ k\Omega)/(11\ 25\ k\Omega + 50\ k\Omega) \approx 0\ 244\ mA; V_L \approx (0\ 24\ mA) \times (50\ k\Omega) \approx 12.2\ V.$

Note that we obtain the same values of V_L as we did in SAQ2 4 and SAQ2 5 using the TEC

SAQ2.9

1. $I_{Norton} = 10\ V/20\ k\Omega = 0\ 5\ mA; R_{Norton} = 20\ k\Omega \parallel 20\ k\Omega = 10\ k\Omega.$
2. $I_L = 0.5\ mA \times 10\ k\Omega/(10\ k\Omega + 5\ k\Omega) = 0\ 333\ mA.$ Therefore, $V_L \approx (0.333\ mA) \times (5\ k\Omega) \approx 1\ 667\ V.$ If we calculate V_L from the actual circuit, we must use the voltage divider equation, the lower resistance of the divider being the parallel combination of 20 kΩ and 5 kΩ

$$V_L = 10\ V \times \frac{(20\ k\Omega \times 5\ k\Omega)/(20\ k\Omega + 5\ k\Omega)}{[(20\ k\Omega \times 5\ k\Omega)/(20\ k\Omega + 5\ k\Omega)] + 20\ k\Omega}$$

$$= 1.667\ V \quad \text{as before!}$$

SAQ2.10

Finding I_{Norton} is easy Short-circuit the lower 20 kΩ and 0.5 mA still flows through the 20 kΩ left. Therefore, $I_{Norton} = 0\ 5\ mA.$ Finding R_{Norton} may confuse you a little. We must open-circuit the current source. So what happens to the top end of the upper 20 kΩ resistor? Nothing. It is floating and no longer forms part of the circuit. Therefore, the resulting resistance between the output terminals is merely 20 kΩ and $R_{Norton} = 20\ k\Omega.$ This means that the NEC is not the same as that in SAQ2 9.

SAQ2.11

The voltage transferred to the amplifier (V_{amp}) would be much smaller than the microphone output voltage (V_{mic}) and would be given by the voltage divider equation

$$V_{amp} = V_{mic} \times 20\ 000\ \Omega/(20\ 000\ \Omega + 50\ 000\ 000\ \Omega) \approx (0.0004)V_{mic}$$

This means that 0.9996 of the microphone output is lost internally, i e. very poor voltage transfer

SAQ2.12

$V_{amp} \approx 0\ 2\ mV.$ Only 0 2 mV reaches the amplifier input

SAQ2.13

The explanation is left to you. However, SAQ2.12 would give a good example. A good voltage source has a low R_{int} (or at least low compared with the load resistance) In SAQ2 12 a good microphone would have $R_{int} \approx 200\ \Omega$ Using this microphone (assuming the same output voltage) with the same amplifier would give

$$V_{amp} = 500 \text{ mV}(20\,000)/(20\,000 + 200) = 495 \text{ mV}$$

Almost all the output reaches the amplifier.

SAQ2.14
High

SAQ2.15
It is assumed that you have drawn the circuit in an attempt to answer the question. The answers are best found by using the current divider equation to find the load current I_L

1   $I_L = 0.1 \text{mA} \times 100 \text{ k}\Omega/(100 \text{ k}\Omega + 80 \text{ k}\Omega) \approx 0.056 \text{ mA}$.
2. $I_L = 0.1 \text{mA} \times 100 \text{ k}\Omega/(100 \text{ k}\Omega + 400 \text{ k}\Omega) = 0.02 \text{ mA}$.
3   $I_L = 0.1 \text{ mA} \times 100\,000\ \Omega/(100\,000\ \Omega + 25\ \Omega) \approx 0.1 \text{ mA} \ (0.0999 \text{ mA})$

SAQ2.16
$$P_{out} = R_L \times [V_s/(R_s + R_L)]^2$$

Therefore

$$P_{out} = (V_s^2) \times [R_L/(R_s + R_L)^2]$$

$dP_{out}/dR_L = d\{(V_s^2) \times [R_L/(R_s + R_L)^2]\}/dR_L$ Taking out V_s as a constant leaves

$$dP_{out}/dR_L = (V_s^2) \times d[R_L/(R_s + R_L)^2]/dR_L$$

From Figure 2.20, P_{out} is a maximum when $dP_{out}/dR_L = 0$ First find dP_{out}/dR_L. To do this we need to substitute and use the product rule. Let $u = R_L$ and $v = (R_s + R_L)^{-2}$. Now

$$d(uv)/dR_L = u(dv/dR_L) + v(du/dR_L)$$

Therefore

$$d(uv)/dR_L = \{R_L \times [-2(R_s + R_L)^{-3}]\} + [(R_s + R_L)^{-2} \times 1]$$

For maximum P_{out} value

$$(V_s^2) \times d(uv)/dR_L = 0$$

Therefore

$$(V_s^2) \times \{R_L \times [-2(R_s + R_L)^{-3}] + [(R_s + R_L)^{-2} \times 1]\} = 0$$

Here we have a product of two terms   One of these terms must be equal to zero for the product to be zero. Since V_s cannot be zero (there would be zero power dissipation if this were true), the only logical outcome is that the other term is zero

$$R_L \times [-2(R_s + R_L)^{-3}] + [(R_s + R_L)^{-2} \times 1] = 0$$

$$R_L \times [-2(R_s + R_L)^{-3}] = -(R_s + R_L)^{-2}$$

$$R_L = -[(R_s + R_L)^{-2} \times (R_s + R_L)^3]/ - 2 = (R_s + R_L)/2$$

$$2 \times R_L = R_s + R_L$$

Therefore, $R_L = R_s$.

SAQ2.17
1. High value $> 10\times$ (600 Ω). Therefore, use $R_L \geq 6$ kΩ.
2. Low value $< 600/10$ Therefore, use $R_L \leq 60$ Ω.
3. For maximum power transfer, $R_L = 600$ Ω

SAQ2.18
This is all to do with good voltage transfer The voltmeter (which will load the output of a circuit it is to measure) needs to have an input resistance hopefully much higher than the output resistance of any circuit voltage it is to measure Since the output resistance of a circuit is unknown, the voltmeter must be made with as large an input resistance as possible (the larger the better) to prevent the loading effect Digital voltmeters often have R_{in} values exceeding $10^{10}\Omega$, whereas CROs typically have $R_{in} = 2$ MΩ.

SAQ2.19
Since the circuit output is loaded significantly by a 1 kΩ load, we can assume that the DVM load of 10^8 Ω has negligible effect Okay, forget the DVM load effect and consider only the circuit and the 1 kΩ load. If the unloaded circuit were converted into a TEC, the output would drop to half its original value if a load equal to the R_{th} were added. Since this is exactly what happens with the 1 kΩ load, we can deduce that the circuit has an R_{th} or output resistance R_{out} of 1 kΩ.

SAQ2.20

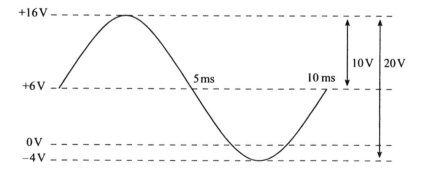

SAQ2.21

1. $V_{pk} = 240 \times (2^{1/2})$ V $= 339.4$ V and $T = (1/50)$ s $= 20$ ms.

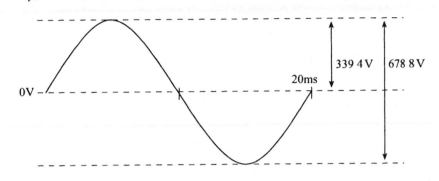

2. $V_{pk} = 12 \times (2^{1/2})$ V $= 17$ V and $T = (1/1500)$ s $= 667$ μs

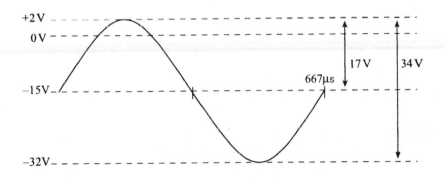

SAQ2.22

$V_{p-p} = 2 \times 10$ V $= 20$ V and $T = (1/1000)$ s $= 1$ ms The centre of the wave is at 10 V.

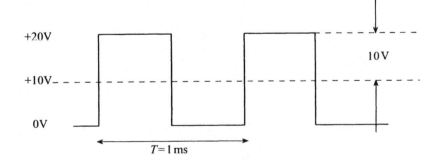

SAQ2.23

This is a 10 V peak-to-peak square wave, with a d c. offset of +7 V and a low level of +2 V (it is wise to add this bit to prevent confusion) at a frequency of 50 Hz.

SAQ2.24

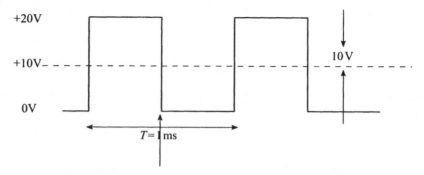

SAQ2.25

$T = 1/5000 = 200\ \mu s$. Therefore, it is high for 50 μs. Duty cycle is the percentage of time that the voltage is high $(50/200) \times 100 = 25\%$, i e. a quarter of the periodic time

SAQ2.26

The high and low times are equal. Therefore, the duty cycle is 50%

SAQ2.27

The waveform rises from −5 V to +5 V in half the periodic time Therefore, because $T = 1$ ms, the gradient is 10 V in 0.5 ms, i.e.

$$10\ V/0.5\ ms = 20\,000\ V\ s^{-1}$$

As the waveform is symmetrical, the falling part of the waveform is equal in magnitude to the rising part, but negative in value It is therefore −20 000 V s⁻¹.

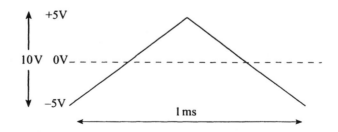

Chapter 3

SAQ3.1

It is difficult to give a general answer here since it really depends on the model you have, but for a standard portable radio-cassette there would be·

- inputs 2 power supplies (mains and battery), aerial, microphone, auxiliary input,
- outputs 2 loudspeaker connections, 1 headphone connection, 1 low level output for another recorder

There are also mechanically controlled inputs such as volume, tone, play and rewind. The details are unimportant: the important part is recognizing an input from an output.

SAQ3.2

4 inputs (power supply (+ and −), inverting and non-inverting) and 1 output (V_{out})

SAQ3.3

$A_{vol} = V_{out}/V_{in}$. Therefore, $A_{vol} = 2/(10 \times 10^{-6}) = 200\,000$ Note no units for the gain

SAQ3.4

Just rearrange the gain equation shown in the answer to the previous question to give $V_{out} = A_{vol} \times V_{in}$

SAQ3.5

We said in Chapter 2 that the input to any circuit is usually represented by a single resistance or impedance Therefore, here we have

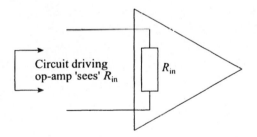

SAQ3.6

The circuit driving the op-amp input would have a finite internal resistance R, and for good voltage transfer to the op-amp input we usually choose R_{in} to be $\geq 10R$ Therefore, the bigger R_{in} is, the better If we do not know the value of R, the simple answer, therefore, is that we want a very large value of R_{in} Many tens of megohms would ensure good voltage from just about any drive circuit

SAQ3.7

If the output of a real op-amp operates as a good voltage source, its internal resistance must be finite but very low A few ohms would be sufficient.

SAQ3.8

We represent the output as a single voltage source and a single resistance (just like a Thevenin equivalent circuit). V_{out} is created by a voltage source $A_{vol} \times V_{in}$ which is referred to earth, as V_{out} is measured with reference to earth. The diagram shows both input and output representations for the op-amp

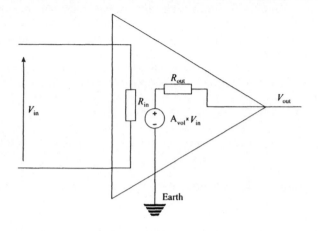

SAQ3.9

The output would be 180° out of phase with the input   The waveforms are not drawn to scale.

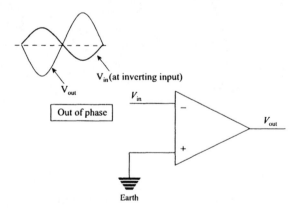

SAQ3.10

Subtracts, feedback.

SAQ3.11

Feedback fraction; β.

SAQ3.12
Positive, larger (than it would have been without the feedback)

SAQ3.13
Negative, smaller (than it would have been without the feedback)

SAQ3.14
Positive

SAQ3.15
Negative

SAQ3.16
$1\,k\Omega$

SAQ3.17
This should be fairly easy. $-R_2/R_1$ or, in other words, the ratio of the feedback resistor to the input resistor

SAQ3.18
The minus sign indicates the inverting action.

SAQ3.19
$A_v = -(1\,k\Omega/1\,k\Omega) = -1$ $V_{out} = A_v \times V_{in} = (-1) \times (2) = -2\,V.$

SAQ3.20
$A_v = -(4\,k\Omega/1\,k\Omega) = -4.$ $V_{out} = A_v \times V_{in} = (-4) \times (2) = -8\,V$

SAQ3.21
Since R_1 remains at $1\,k\Omega$, R_i of both circuits is $1\,k\Omega$

SAQ3.22
As $R_i = 20\,k\Omega$, $R_1 = 20\,k\Omega$ Since $R_2/R_1 = A_v = -20$, $R_2 = 20 \times R_1 = 20 \times 20\,k\Omega = 400\,k\Omega$

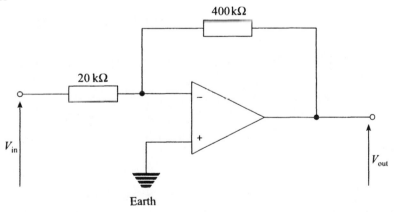

SAQ3.23

$A_v = 1 + (R_2/R_1) = 1 + (10/2) = 6.$

SAQ3.24

$A_v = 1 + (R_2/R_1) = 1 + (5/0\ 5) = 11$ $V_{out} = A_v \times V_{in} = 11 \times (500\ mV) = 5\ 5\ V.$

SAQ3.25

For good voltage transfer, the load circuit must have a high resistance Since the non-inverter has an input resistance which is extremely large (ideally, we assume it is infinitely large), we therefore have *good* voltage transfer

SAQ3.26

Remember here that we are considering the same circuit as that in Figure 3.11, with the added complication that the voltage follower has finite input and output resistances The input circuit forms a voltage divider between the 1000 Ω source resistance and the 1 MΩ voltage follower input resistance Using the potential divider equation, we have

$$V = 1\ V \times 1\ M\Omega/(1\ M\Omega + 1000\ \Omega) = 0.999\ V$$

The voltage follower has a gain of 1; therefore, the output voltage V_{out} is 0.999 V. This output is divided between the voltage follower's output resistance and the 10 Ω load. Therefore, using the potential divider equation again to find the voltage V_L across the 10 Ω load, we have

$$V_L = 0\ 991\ V \times 10\,\Omega/(1\,\Omega + 10\,\Omega) = 0.908\ V$$

Note that nearly all of the 1 V source reaches the 10 Ω load.

SAQ3.27

1. Worse
2 Worse
3. Better. In fact, the transfer would be good enough not to necessitate using the voltage follower circuit at all
4 Worse
5 No different
6. Worse, although it would have to be increased to a very large value for any deterioration to be noticeable
7. Better.
8 Better
9 Worse This would make the voltage follower a poorer voltage source.

Chapter 4

SAQ4.1

The minus sign indicates inversion For this circuit it indicates that a positive input current (taken to be flowing *into* the op-amp) will produce a negative V_{out} value

SAQ4.2

$0\,\Omega$.

SAQ4.3

Using $V_{out} = -R_f \times I_{in}$, we get

$$V_{out} = -(2 \text{ k}\Omega) \times (2\ 5 \text{ mA}) = -5 \text{ V}.$$

SAQ4.4

In this circuit, input current flow is taken as positive *into* the op-amp, whereas here we have input current flow in the opposite direction, which is therefore negative. Therefore, $I_{in} = -5 \text{ mA}$

$$V_{out} = -R_f \times I_{in} = -(0.1 \text{ k}\Omega) \times (-5 \text{ mA}) = +0\ 5 \text{ V}$$

SAQ4.5

The R_{in} value for each input is merely $1 \text{ k}\Omega$. the size of each of the resistors in the input circuit Note that the neighbouring voltage sources do not affect R_{in} seen by each input. This is because of the virtual earth at the summing point.

SAQ4.6

With reference to Figure 4 4, since we now have four inputs we would add another input V_4 into a resistance R_4. We can now call the feedback resistance R_f For each input to have an R_{in} of $10 \text{ k}\Omega$, we set $R_1 - R_2 - R_3 - R_4 = 10 \text{ k}\Omega$ If the amplifier is to give equal weighting to all four inputs, then we must set $R_1 = R_2 = R_3 = R_4$. But you can see that we already have this condition! This should tell us something about this circuit: namely, that if the inputs have equal weighting they must automatically also have equal input resistance. We now finish this off by setting a value for R_4. Make $R_4 = 10 \text{ k}\Omega$ and we have a unity gain circuit V_{out} would be given by

$$V_{out} = -(V_1 + V_2 + V_3 + V_4) \times R_f/R$$

$$= -(0.5 \text{ V} + 0\ 03 \text{ V} + 4\ 1 \text{ V} + 2.04 \text{ V}) \times (10 \text{ k}\Omega/10 \text{ k}\Omega)$$

$$= -6.67 \text{ V}$$

SAQ4.7

A brief description only is required here The circuit is a two-input voltage adder with all resistors variable If the feedback resistance is R_f and the inputs are V_1 and V_2 into resistances of R_1 and R_2 respectively, reducing R_1 will increase the gain of the V_1 signal, reducing R_2 will increase the gain of the V_2 signal and increasing R_f will increase the gain of both signals together

SAQ4.8

See diagram on page 294

SAQ4.9

See diagram on page 294

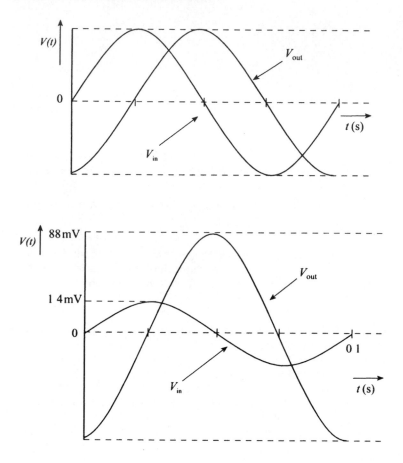

SAQ4.10

The waveform diagram will be similar to that for the previous question, with different input and output amplitudes $V_{in(peak)}$ is given as 10 V Since we know that V_{in} is a sine wave, we can determine V_{out} using the $-(R/Z)$ gain method Where $R = 20$ kΩ and $Z = 1/(2 \times \pi \times f \times C) = 1/[2 \times \pi \times 1000 \times (0\ 01\mu F)] = 15\ 92$ kΩ

$$A_v = -R/Z = -20\,000/15\,920 = -1\ 26$$

$$V_{out} = A_v \times V_{in} = -1.26 \times 10 = -12.6 \text{ V}$$

SAQ4.11

Since $A_v = -R/Z$ and $Z = 1/(2 \times \pi \times f \times C)$, we can see that if f increases, Z decreases, causing the voltage gain to increase. Therefore, higher frequencies produce more gain. Using $V_{out} = \{-R/[1/(2 \times \pi \times f \times C)]\} \times V_{in}$, at $f = 2$ kHz

$$V_{out} = \{-R/[1/(2 \times \pi \times f \times C)]\} \times V_{in} = -25 \text{ V}$$

at $f = 10$ kHz

$$V_{out} = \{-R/[1/(2 \times \pi \times f \times C)]\} \times V_{in} = -126 \text{ V}$$

at $f = 20$ kHz

$$V_{out} = \{-R/[1/(2 \times \pi \times f \times C)]\} \times V_{in} = -251 \text{ V}$$

and at $f = 50$ kHz

$$V_{out} = \{-R/[1/(2 \times \pi \times f \times C)]\} \times V_{in} = -628 \text{ V}$$

These V_{out} values would be impossible to reach for a real op-amp and we would obtain considerable distortion of the output instead. However, this is merely a limitation of the real device and should not prevent us from understanding that the output gets bigger as the frequency increases.

SAQ4.12
If we assume that the required input signal is a low frequency sine wave which contains a high frequency sine wave interference signal (such as you might see in an audio amplifier which is unfortunately picking up an interfering radio broadcast from a nearby transmitter), since the interference signal is a high frequency, the differentiator amplifies its differential with a far higher gain than that of the required signal The resulting differentiated output contains a massive interference signal which completely swamps the small low frequency signal.

SAQ4.13
We require $CR = 1$ One choice for this would be $C = 1 \mu F$ and $R = 1$ MΩ, but of course there are others. For example, if $R = 10$ MΩ, $C = 100$ nF. The waveform diagram looks like

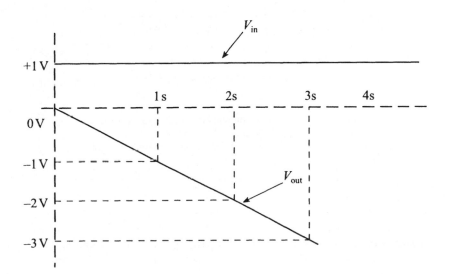

SAQ4.14

$$V_{out} = -1/CR \int_{t_1}^{t_2} V_{in} \, dt$$

$CR = (1 \, \mu F) \times (100 \, k\Omega) = 0.1 \, s$ and therefore $1/CR = 10 \, s^{-1}$ Therefore, integrating from time zero to 1 ms gives

$$V_{out} = -(1/CR) \times [V_{in} \times t]_{t_1}^{t_2}$$

(where $t_1 = 0$ and $t_2 = 1$ ms)

$$= -(10) \times [V_{in} \times t]_0^{1ms} = -10[(1 \, V \times 10^{-3}) - (0)] = -0 \, 01 \, V$$

Since $V_{in} = 0$ after 1 ms, the integral will also be zero and can be ignored Alternatively, at 1 ms the area (and therefore the integral) traced out by the input signal is a rectangle 1 V high and 1 ms long After 1 ms, $V_{in} = 0$ and therefore traces out no more area

$$\text{area} = (1 \, V) \times (1 \, ms) = 1 \times 10^{-3} \, V \, s$$

Remembering now to multiply by the gain factor $(1/CR)$, we have

$$V_{out} = -10 \times (1 \times 10^{-3}) = -0.01 \, V$$

Either method yields the correct answer, to give a waveform diagram looking like

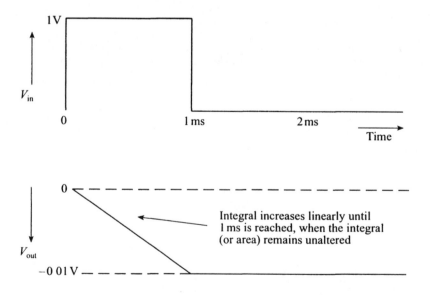

SAQ4.15
Since $V_{p-p} = 5 \, V$ and the d.c offset is 0, the waveform rises from 0 to +2 5 V for half its period and falls to −2.5 V for the other half, sweeping out equal (and opposite) areas each time. We can therefore intuitively predict that the output will be a triangle wave, with a V_{p-p} value given by

$$V_{p-p} = -(1/CR) \times [V_{in} \times t]_{t_1}^{t_2} = -(10) \times [V_{in} \times t]_0^{1ms} = -10[(2\ 5 \times 10^{-3}) - (0)]$$
$$= -0.025\ V \text{ or } 25\ mV.$$

SAQ4.16
If f in the answer to the previous question were doubled, T (and therefore $T/2$) would be halved, making $t_2 = 0.5$ ms. The resulting output would be

$$V_{p-p} = -(10) \times (V_{in} \times t)_0^{0\ 5ms} = -10\{[2.5\ V \times (0\ 5 \times 10^{-3})] = (0)\}$$
$$= -0.0125\ V \text{ or } 12.5\ mV.$$

SAQ4.17
Using CMRR $= 20 \log_{10}(A_{v(diff)}/A_{v(cm)}) = 90$ dB, we can say $90/20 = \log_{10}(A_{v(diff)}/A_{v(cm)})$ Therefore, $4.5 = \log_{10}(A_{v(diff)}/A_{v(cm)})$ and finally, from the law of logs, $10^{4\ 5} = A_{v(diff)}/A_{v(cm)}$, i e $A_{v(diff)}/A_{v(cm)} = 31\ 623$. In other words, the difference gain is 31 623 times bigger than the common mode gain.

SAQ4.18
From Figure 4 11, $A_{v(diff)} = 100$ MΩ/500 k$\Omega = 200$ (directly from the resistance values, note the symmetry).

$A_{v(cm)} = V_{out(cm)}/V_{in(cm)}$ and Figure 4 11 shows a labelled common mode input $V_{in(rms)}$ and its resulting output $V_{out(rms)}$. Using V_{rms} values is fine so long as we do not mix V_p and V_{p-p} values in the same expression. Therefore

$$A_{v(cm)} = V_{out(cm)}/V_{in(om)} = 500\ mV/20\ V = 0.025$$

Therefore, CMRR $= A_{v(diff)}/A_{v(cm)} = 200/0.025 = 8000$ and CMRR $= 20 \log_{10}(A_{v(diff)}/A_{v(cm)}) = 20 \log_{10}(8000) = 20 \times 3.90 = 78.1$ dB.

Chapter 5

SAQ5.1
Using GBWP$_{(741)} = 1$ MHz, $A_v = -20$ and GBWP $= A_v \times$ BW, we have BW $=$ GBWP$/A_v$ $= 10^6/20 = 50$ kHz (We ignore the minus sign as it is meaningless in this context)

SAQ5.2
From Figure 5 5 we can estimate (and verify from the GBWP expression) that $A_{vol(741)}$ is 200 at 5 kHz, 20 at 50 kHz and 2 at 500 kHz.

SAQ5.3
At 10 kHz, $A_{vol} = 100$. It would therefore be sensible to keep A_v values to $\leq 100/10$, i e. a sensible maximum of A_v is about -10.

SAQ5.4
Using these resistances, the gain is calculated at $A_v = -1000$ kΩ/1 k$\Omega = -1000$. At 5 kHz, $A_{vol} = 200$ Therefore, we obviously cannot successfully achieve an A_v level of >200 We would be better to redesign the circuit with a gain of about 20

SAQ5.5
Unit gain frequency, f_{unity}

SAQ5.6
$I_{bias} = (10\,nA + 20\,nA)/2 = 15\,nA$, $I_{in(off)} = |\,10\,nA - 20\,nA\,| = 10\,nA$.

SAQ5.7
Intuitively, $I_{bias} = 100\,nA$ is an average of I_{b1} and I_{b2}, which differ by 20 nA The input current values are therefore 110 nA and 90 nA.
A more rigorous method follows, but remember that being intuitive (and quick) is a great asset for an engineer.

$$|I_{b1} - I_{b2}| = 20\ nA$$

Therefore

$$I_{b1} = 20\ nA + I_{b2}$$

Substituting this into

$$(I_{b1} + I_{b2})/2 = 100\ nA$$

gives

$$(20\ nA + I_{b2} + I_{b2})/2 = 100\ nA$$

i e.

$$I_{b2} = 100\ nA - 10\ nA = 90\ nA$$

Therefore

$$I_{b1} = 20\ nA + 90\ nA = 110\ nA.$$

SAQ5.8
We do not know.

SAQ5.9
The centre or zero-crossing point.

SAQ5.10
A 30 kHz triangle wave has a period T of $1/30\,000 = 33\,\mu s$ Therefore, a $20\,V_{pk-pk}$ wave rises (or falls) through 20 V in half the periodic time· 20 V in 16 5 μs. The maximum rate of change of input voltage into the follower circuit is therefore

$$dV/dt = (20/16.5)\ V\mu s^{-1} = 1.2\ V\mu s^{-1}$$

Since the SR of the circuit is only $1\,V\,\mu s^{-1}$, the output cannot change quickly enough and is therefore slew rate limited.

SAQ5.11
A $10\,V_{pk-pk}$, 20 kHz sine wave has a v_{pk} of 5 V and $\omega = 2\pi f \approx 125\,664\ rad\,s^{-1}$. We therefore describe the waveform as $5\sin(\omega t)$ We now need to find the maximum

dv/dt of the wave, which means differentiating the function. We therefore have

$$d(5 \sin \omega t)/dt = 5\omega \times \cos \omega t$$

We know that for a sine wave this is a maximum value at $t = 0$. Therefore

$$(dv/dt)_{max} = 5\omega \times \cos 0$$

Using $\cos 0 = 1$ and $\omega = 125\,664$, we have

$$(dv/dt)_{max} = (5 \times 125\,664) = 628\,320 \text{ V s}^{-1}$$

$$\approx 0\,63 \text{ V } \mu s^{-1}$$

Since the SR of the circuit is only $0\,5$ V μs^{-1}, the output cannot change quickly enough and is therefore slew rate limited

SAQ5.12
This question is concerned merely with the output saturation levels of the op-amp Assume that the output saturation levels will be approximately $+/-9$ V

$$V_{in} = 1 V_{rms} = 1\,414 V_{pk}$$

The gain required to take the output to saturation is

$$A_v = 9/1\,414 \approx 6\,4$$

Therefore, to avoid clipping it would be best to reduce the maximum gain to $A_v = 6$

SAQ5.13
For an op-amp with an I_{bias} of 200 nA and $V_{in(off)} = 1$ mV in an Inverter with $A_v = -500$, we need an input resistance of at least 100 kΩ. To determine the size of the output offset voltage $V_{o(os)}$, we must choose the R_1 and R_2 values We will make $R_1 = 100$ kΩ and $R_2 = 50$ MΩ, to give the required A_v and to satisfy the R_{in} requirement. Now

$$R_1 \| R_2 = 100 \text{ k}\Omega \times 50 \text{ M}\Omega/(100 \text{ k}\Omega + 50 \text{ M}\Omega) = 99\,8 \text{ k}\Omega$$

Therefore, the voltage drop is

$$V_{-os} = 200 \text{ nA} \times 99\,8 \text{ k}\Omega = 20 \text{ mV}$$

and including $V_{in(off)}$ gives

$$V_{i(os)} = 1 \text{ mV} + 20 \text{ mV} = 21 \text{ mV}$$

This is increased by the gain to give

$$V_{o(os)} = 21 \text{ mV} \times (-500) = 10\,5 \text{ V}$$

SAQ5.14
Initially we will choose R_1 and R_2 values to give $A_v = -200$ Since no specification for R_{in} is given, we will set it at 1 kΩ by using $R_1 = 1$ kΩ and $R_2 = 200$ kΩ We are not given $V_{in(off)}$, therefore we must ignore it in any calculation. Now

$$R_1 \| R_2 = 1 \text{ k}\Omega \times 200 \text{ k}\Omega/(1 \text{ k}\Omega + 200 \text{ k}\Omega) = 0\,995 \text{ k}\Omega$$

Therefore, the voltage drop is

$$V_{-(os)} = 400 \text{ nA} \times 0.995 \text{ k}\Omega = 398 \ \mu\text{V}$$

Assuming $V_{in(off)} = 0$, we have

$$V_{i(os)} = 0 + 398 \ \mu\text{V} = 398 \ \mu\text{V}$$

This is increased by the gain to give

$$V_{o(os)} = 398 \ \mu\text{V} \times (-200) = 79.6 \text{ mV}$$

This is not much of a problem, but in an attempt to reduce it we would introduce an extra resistance (see Figure 5 8), $R_3 \approx 0$ 995 kΩ. A suitable value would be $R_3 =$ 1 kΩ.

SAQ5.15
1. Your drawing of the circuit should look like Figure 5.9, where $R_3 = 25$ kΩ and $R_1 =$ 57 kΩ
2. We need to find R_2 to give an A_v of 20 Using $A_v = 1 + R_1/R_2$, we have $R_2 = R_1/(20 - 1) = 3$ kΩ.
3. The offset at the non-inverting input is $V_{+(os)} = I_{bias} \times R_s$, where $R_s = 25$ kΩ, giving $V_{+(os)} = I_{bias} \times 25$ kΩ, whereas the offset at the inverting input is $V_{-(os)} = I_{bias} \times (R_1 \parallel R_2)$, where $R_1 \parallel R_2 = (57 \text{ k}\Omega \times 3 \text{ k}\Omega)/(57 \text{ k}\Omega + 3 \text{ k}\Omega) = 2$ 85 kΩ Therefore, $V_{-(os)} = I_{bias} \times 2$ 85 kΩ. From the two expressions for the offsets we can see that $V_{+(os)} > V_{-(os)}$.
4 Using $V_{i(os)} = V_{in(off)} + |(V_{+(os)} - V_{-(os)})|$ gives the input offset as

$$V_{i(os)} = 2 \text{ mV} + |(12.5 \text{ mV} - 1.5 \text{ mV})| = 2 \text{ mV} + 11 \text{ mV} = 13 \text{ mV}$$

The output offset is given by

$$V_{o(os)} = V_{i(os)} \times A_{vcl}$$
$$= 13 \text{ mV} \times 20 = 260 \text{ mV}.$$

Chapter 6

SAQ6.1
1 $V_+ = 15 \times (12 \text{ k}\Omega)/(12 \text{ k}\Omega + 100 \text{ k}\Omega) \approx 1$ 61 V ($V_+ < 7.5$ V, therefore V_{out} switches to about -14 V.)
2 $V_+ = 15 \times (12 \text{ k}\Omega)/(12 \text{ k}\Omega + 2 \text{ k}\Omega) \approx 12$ 86 V ($V_+ > 7.5$ V, therefore V_{out} switches to about $+14$ V and sets off the alarm.)

SAQ6.2
Interchanging the thermistor and the 12 kΩ resistor would give cold and hot V_+ values of 13 4 V and 2.1 V respectively The effect would be to set the alarm off when the thermistor is cold (i e. the opposite of the previous case).

SAQ6.3

We want to design a delay timer to switch an op-amp output to +14 V after 45 s The circuit looks like Figure 6 3 We know that $t = 45$ s, therefore we set $0.7 \times CR = 45$ s Therefore, $CR = 45/0$ 7 s. Choose $C = 50\ \mu F$. Then we have $R = 45/(0\ 7 \times 50\mu F) \approx 1.29\ M\Omega$ (It is wise to make this a variable resistance)

SAQ6.4

If we reduce V_{ref}, the capacitor voltage rises exponentially and would therefore reach the reduced value earlier, causing a reduction in time delay.

SAQ6.5

Increasing V_{ref} towards V_{target} (i e. +15 V) increases the time delay. Theoretically, the capacitor takes an infinite time to reach its target, therefore, a V_{ref} close to +15 V would give very long delays. However, because of the asymptotic nature of the exponential curve at high t values, the exact delay becomes very difficult to predict

SAQ6.6

Doubling C will merely double the delay time to 90 s (i e 2×45 s).

SAQ6.7

Referring to Figure 6 4, there are various ways of doing this. All involve generating +2 V at the inverting input of op-amp 1 and −2 V at the non-inverting input of op-amp 2 The simplest way is to use the voltage divider shown in Figure 6 4, but remember that the voltage drop across R_2 is $[(+2) - (-2)] = 4$ V Each resistor must therefore drop 4 V. This means driving the divider circuit with +6 V (instead of +15 V) and −6 V (instead of 0 V)

SAQ6.8

For this situation we want +5 V at the inverting input of op-amp 1 and 0 V at the non-inverting input of op-amp 2 Again, using the voltage divider in Figure 6.4, the voltage drop across R_2 is $[(+5) - (0)] = 5$ V Each resistor must therefore drop 5 V This means driving the divider circuit with +10 V (instead of +15 V) and −5 V (instead of 0 V)

SAQ6.9

We calculated in answer to SAQ6.1 that the cold and hot divider output voltages were 1 61 V and 12 86 V respectively. This voltage range is quite sufficient to set off both alarms at each extreme of temperature However, the question asks us to find the thermistor values required to *just* trigger each alarm, i e the $R_{thermistor}$ value which would give slightly less than +5 V and slightly more than +10 V First, we need to find the $R_{thermistor}$ value which would give exactly these voltages. Therefore, for 5 V

$$+5 = 15 \times (12\ k\Omega)/(R_{thermistor} + 12\ k\Omega)$$

which gives

$$R_{thermistor} = [(12 \times 15)/5] - 12$$

$$= 24 \text{ k}\Omega$$

For 10 V

$$+10 = 15 \times (12 \text{ k}\Omega)/(R_{thermistor} + 12 \text{ k}\Omega)$$

which gives

$$R_{thermistor} = [(12 \times 15)/10] - 12$$

$$= 6 \text{ k}\Omega$$

Since the thermistor is the upper resistance of the divider·
1   the over-temperature alarm will be set when $R_{thermistor} < 6 \text{ k}\Omega$,
2.   the under-temperature alarm will be set when $R_{thermistor} > 24 \text{ k}\Omega$.

SAQ6.10
Here we have equal resistors in the divider, therefore, the reference voltage is merely half the op-amp output value: $V_{ref} = +/-7 \text{ V}$, i e $+/-(14/2)$.

SAQ6.11
$[(+7 \text{ V}) - (-7 \text{ V})] = 14 \text{ V}$

SAQ6.12
Using the circuit component values of Figure 6.6 with a $+/-9$ V power supply means that the op-amp output saturation levels will be approximately $+/-8$ V Therefore, V_{ref} will be

$$V_{ref} = (+/- 8 \text{ V}) \times (1 \text{ k}\Omega)/(1 \text{ k}\Omega + 13 \text{ k}\Omega) = +/- 0.57 \text{ V}$$

$$\text{Hysteresis} = (+0 \ 57 \text{ V}) - (-0 \ 57 \text{ V}) = 1.14 \text{ V}$$

SAQ6.13
We would need to increase the difference in the on/off trigger levels to prevent multiple switching Therefore, we would increase the hysteresis

SAQ6.14
We have equal resistances for R_1 and R_2; therefore

$$f \approx [2.2CR]^{-1} \approx [(2 \ 2) \times (0.01\mu F) \times (1 \text{ k}\Omega)]^{-1} \approx 46 \text{ kHz}$$

$$f = [2CR \log_e(3)]^{-1} = [(2) \times (0.01\mu F) \times (1 \text{ k}\Omega) \times (1.0986)]^{-1} = 45.51 \text{ kHz}.$$

SAQ6.15
1   Increasing R will reduce f.
2   Decreasing C will increase f
3.   Increasing the op-amp power supply voltage presents us with a dilemma. The V_{ref} values will increase, tempting us into thinking that C will take longer to charge up to the increased reference voltage. However, the CR part of the circuit will be charging towards an increased target voltage also! Since the time taken to charge to a specific

fraction of the target voltage is always fixed (by CR), and since the reference voltage is always a fixed fraction of the target voltage (fixed by R_1 and R_2), the output frequency remains almost unchanged by power supply variations

4   Reducing R_2 will result in a reduced V_{ref} and C will charge to this new value more quickly. The frequency therefore increases.

5   Reducing R_1 will result in an increased V_{ref} and C will charge to this new value more slowly The frequency therefore decreases.

SAQ6.16
One is merely a divided down version of the other and is therefore in phase with it.

SAQ6.17
We will assume that the resulting output saturation values are +/−9 V. We are given all the values for the equation for T. Thus

$$T = 2CR \log_e(1 + 2R_2/R_1)$$

Therefore

$$T = (2) \times (100 \text{ nF}) \times (5 \text{ k}\Omega) \times [\log_e\{1 + [2 \times (40 \text{ k}\Omega/10 \text{ k}\Omega)]\} = 2.197 \times 10^{-3} \text{ s}$$

$$f = (2\ 197 \times 10^{-3})^{-1} = 455 \text{ Hz}$$

The V_{out} square wave will oscillate from +9 V to −9 V The voltage across R_2 is a divided down version of V_{out}

$$V_{R2} = (+/-9) \times (40 \text{ k}\Omega)/(40 \text{ k}\Omega + 10 \text{ k}\Omega) = +/-7.2 \text{ V}.$$

SAQ6.18
We are looking for a charging curve that is as straight as possible. To achieve this we would like to use only the very early parts of the charging curve where the exponential curve appears almost linear. The CR value governs the time for charging to a specific value, the shape of the capacitor waveform will be the same for any CR charging to the same value. Therefore, a large CR is of no use. For the same reason a small CR is of no use. For reasons mentioned in the answer to SAQ6.15, the power supply has minimal effect on circuit action and so a large power supply is of no use Likewise, a small power supply is of no use. Large R_2 increases the V_{ref} value and the time for C to charge. C will charge to a value higher up its exponential curve, which means that the waveform becomes rather curved and hence of no use as a triangle wave However, a small R_2 decreases the V_{ref} value and the time for C to charge. C will charge to a value lower down its exponential curve (nearer the start), which means the waveform is less curved and hence generates a useful triangle wave.

SAQ6.19
Keep things simple and use equal values of R_1 and R_2. $R_1 = R_2 = 10 \text{ k}\Omega$ will do nicely For $f = 5$ kHz, we now use the approximation $f = (2.2CR)^{-1}$ Therefore, $CR = (2.2f)^{-1} = (2\ 2 \times 5 \text{ kHz})^{-1} = 90\ 9\ \mu\text{s}$. Choose a small C value like $C = 0.01\ \mu\text{F}$. Therefore, $R = 90\ 9\ \mu\text{s}/0\ 01\ \mu\text{F} = 9\ 09 \text{ k}\Omega$.

SAQ6.20

In the answer to SAQ6.19 we used an approximate equation but, as 9.09 kΩ resistors are hard to buy, it would be wise to use a 10 kΩ variable resistor and to adjust it to give the exact frequency required. This technique is a useful one in circuits of this type.

Chapter 7

SAQ7.1

The voltage supply is d c. Therefore, we use R_D and the Ge diode action takes up 0 3 V. We can see, therefore, that

$$V_{circuit} = V_{in} - 0.3 = 10 - 0.3 = 9.7 \text{ V}$$

The circuit current (I) is therefore

$$I = V_{circuit}/(R + R_D) = 9\,7 \text{ V}/(10\,000 + 40)\Omega = 966 \text{ }\mu\text{A}.$$

SAQ7.2

In this case, $V_{in} = 5$ V. Therefore

$$V_{circuit} = V_{in} - 0.6 = 5 - 0\,6 = 4.4 \text{ V}$$

We are told that $R_D = 50 \text{ }\Omega$ and $I = 50$ mA (0 05 A). Therefore, rearranging

$$I = V_{circuit}/(R + R_D)$$

gives

$$R = (V_{circuit}/I) - R_D = (4.4/0.05) - 50 = 38\Omega.$$

SAQ7.3

Replacing Ge by Si in a diode circuit would cause an increase in V_T and, hence, the circuit voltage $V_{circuit}$ across the other components to decrease. The circuit current would therefore fall.

SAQ7.4

As the temperature T drops, V_T rises to V_{T1} and I_s drops to I_{s1}. A drop in 20°C means $dT = 20$ and the number of 10°C intervals in dT is

$$d = dT/10 = 20/10 = 2$$

This time we have a halving effect, i e. I_s is halved twice. Therefore

$$I_{s1} = 100 \text{ nA} \times 0.5^2$$

$$= 100 \text{ nA} \times 0.25 = 25 \text{ nA}.$$

SAQ7.5

$V_T = 300$ mV and rises to V_{T1} with a temperature drop from 80°C to 10°C at a rate of 2 mV per °C. Therefore

$$V_{T1} = 300 \text{ mV} + [(80 - 10) \times 2] \text{ mV} = 300 \text{ mV} + 140 \text{ mV} = 440 \text{ mV}$$

SAQ7.6

The gradient of the diode curve is found from

$$(\text{change in } I)/(\text{change in } v) = dI/dV \ (\text{equivalent to } I/V) = 1/R$$

from Ohm's law Therefore

$$1/\text{gradient} = dV/dI = R$$

SAQ7.7

At higher currents the diode curve becomes more vertical, indicating a reduction in resistance. In other words, from the diode curve we can see that a current can increase a large amount, while V increases only slightly Since $R_D = V/I$, this results in a reduction of R_D.

SAQ7.8

Using $r_d = 26/I$ (where I is in mA), we have $r_d = 26/1 = 26 \ \Omega$.

SAQ7.9

Using $I_D \approx I_s \times \exp(V_D/nK)$, where $K = 0 \ 026$ V at room temperature and $n = 1$, we have

$$I_D \approx I_s \times \exp[V_D/0.026]$$

Differentiating this expression with respect to V_D, we get dI_D/dV_D, which is the reciprocal of resistance (in this case of dynamic resistance) $1/r_d$ The differentiated expression is therefore

$$dI_D/dV_D = 1/r_d = (1/0.026 \times I_s \times \exp(V_D/0.026)$$

Note that this expression contains the term $I_s \times \exp(V_D/0.026)$, which is already defined as I_D. The expression therefore simplifies to

$$1/r_d = (1/0 \ 026) \times I_D$$

which give us

$$r_d = 0 \ 026/I_D$$

We can now simply write the steady diode current as I, and we have the result

$$r_d = 0.026/I$$

If we wish, we can now convert 0.026 V into 26 mV and current I from A to mA, and we have $r_d = 26/I$, where I is in mA

SAQ7.10

The forward bias across an Si diode can never rise above its V_T value. Therefore, in this case the diode must have blown, causing an open circuit, where the circuit current $I = 0$ and all the supply voltage appears across the open-circuited component, which is the diode.

SAQ7.11
The current flow through the LED will be reduced and this will reduce the brightness of the LED

SAQ7.12
We do not use the I_{max} value because at this current the component is likely to have a shorter lifetime. We always avoid, if possible, running devices at the extreme top end of their specification.

SAQ7.13
The maximum reverse bias voltage across D1 is limited by the LED forward bias of 1.8 V The maximum reverse bias voltage across the LED is limited by the forward bias across D1 of 0 6 V

SAQ7.14
During the negative half-cycle, D1 is forward biased at 0.6 V The maximum voltage from the supply is $V_p = 358$ V (calculated in Example 7.6) and the current I is limited by the previously calculated resistance $R = 8.91$ kΩ. Therefore

$$I = (358 - 0\,6)/8910 = 40.1 \text{ mA}$$

(just slightly more than the LED current).

SAQ7.15
We would choose a diode with $I_{max} > 40$ mA, $I_{max} = 0.1$ A would do!

SAQ7.16
The circuit used would be that in Figure 7.11, where the input voltage would be a 9 V battery rather than the a.c. supply shown in Figure 7.11. We would choose green for correct battery polarity and red for incorrect. With reference to the circuit in Figure 7.11, this means green for LED1 and red for LED2. The LED which will carry the larger current is the one with smaller V_T, which is the red one with $V_T \approx$ 1.6 V. We choose a current flow below the I_{max} value Choose $I = 20$ mA To find R, we say

$$R = (9 - 1.6) \text{ V}/20 \text{ mA} = 0.37 \text{ k}\Omega = 370\Omega$$

When the battery is connected correctly the green LED lights, dropping 2.2 V from the circuit. Therefore

$$I = (9 - 2\,2) \text{ V}/0.37 \text{ k}\Omega = 18\,4 \text{ mA}.$$

SAQ7.17
$C = I_L/(f_{out} \times v_r)$ and we can rewrite this as $v_r = I_L/(C \times f_{out})$. Therefore:
1. v_r decreases;
2. v_r increases;
3   v_r decreases.

SAQ7.18

Using $C = I_L/(f_{out} \times v_r)$

$$v_r = 5\% \text{ of } 50V = (5/100) \times 50 \text{ V} = 2.5 \text{ V}$$

$$I_L = 0.1 \text{ A}$$

In the UK, mains f_{in} is 50 Hz. Therefore, for the rectifier $f_{out} = 2 \times f_{in} = 2 \times 50 = 100$ Hz and

$$C = 0.1/100 \times (2.5) = 400 \times 10^{-6} \text{ F} = 40\mu F$$

Using $V_{out} = (V_p - 1 2)$, where V_p Is the peak rectifier input voltage and the required rectifier output voltage V_{out} is 50 V, we have

$$V_p = V_{out} + 1.2 = 50 + 1 2 = 51.2 \text{ V}$$

Since $V_{rms} = V_p \times (0 7071)$, the required r m.s rectifier input voltage $V_{rms} = (51.2) \times (0 7071) = 36 2 V_{rms}$

SAQ7.19

Using $C = (V_L/R_L)/(f_{out} \times v_r)$

$$v_r = 10\% \text{ of } 25 \text{ V} = (10/100) \times 25 \text{ V} = 2.5 \text{ V}$$

$$V_L/R_L = 25 \text{ V}/100 \text{ } \Omega = 0.25 \text{ A}$$

In the USA, mains f_{in} is 60 Hz Therefore, for the rectifier $f_{out} = 2 \times f_{in} = 2 \times 60 = 120$ Hz and

$$C = (0.25)/(120 \times 2.5) = 833 \times 10^{-6} \text{ F} = 833\mu F$$

Using $V_{out} = (V_p - 1 2)$, where V_p is the peak rectifier input voltage and the required rectifier output voltage V_{out} is 25 V, we have

$$V_p = V_{out} + 1.2 = 25 + 1.2 = 26.2 \text{ V}$$

Since $V_{rms} = V_p \times (0 7071)$, the required r m.s. rectifier input voltage $V_{rms} = (26 2) \times (0 7071) = 18 5 V_{rms}$.

SAQ7.20

1. 100 Hz.
2   120 Hz.

SAQ7.21

Find a suitable R value and zener power rating for the zener voltage regulator in Figure 7 29(a) $R_L = 600 \text{ } \Omega$. Then find R. We are given $V_L = V_z = 3 3$ V and $V_{in} = 10$ V. Therefore, from

$$R = [(V_{in}/V_z) - 1]R_L$$

we have

$$R = [(10/3 3) - 1] \times 600 \text{ } \Omega = 1.22 \text{ k}\Omega$$

Because real resistors have a +/– percentage error, we ensure that the zener will conduct with this resistance by choosing the next preferred resistor value below this: say, $R = 1.2 \text{ k}\Omega$. To find the power rating, if R_L is disconnected we have

$$I_z = I_R = (V_{in} - V_{out})/R$$

Therefore

$$I_z = (10 - 3.3)/1\,2 \text{ k}\Omega = 5.58 \text{ mA}$$

Hence

$$P_z = V_z \times I_z = (3.3 \text{ V}) \times (5\,58 \text{ mA}) = 18.4 \text{ mW}$$

We would therefore use any available power rating larger than this.

SAQ7.22

This is much like the answer to SAQ7 21, with the addition that R_L is not given directly and is found from $R_L = V_L / I_L = 12 \text{ V}/500 \text{ mA} = 24\,\Omega$. R is then found to be any value less than 76 Ω. Therefore, choose $R = 75\,\Omega$ and the power rating comes out at any value where $P_z > 6.1 \text{ W}$.

SAQ7.23

The solution technique again is almost as that of SAQ7.21. This time, however, we have the added complexity of the load current varying. We are concerned with the smallest R_L value which may be used, which means when I_L is its largest. Therefore, we use $I_L = 500 \text{ mA}$

$$R_L = V_L/I_L = 3\,9 \text{ V}/500 \text{ mA} = 7.8\,\Omega$$

This gives us $R < 16.2\,\Omega$. Therefore, choose $R = 16\,\Omega$ and $P_z > 1.97 \text{ W}$

SAQ7.24

If the zener is running hot it means that it is carrying too much current (i.e. its power rating is being exceeded) The current through R is shared between R_L and the zener. Therefore, to reduce the zener current we increase the load current by reducing R_L

SAQ7.25

The voltage across the load (1 kΩ) would, by voltage divider action, be

$$V_L = 20 \times (1 \text{ k}\Omega)/(1 \text{ k}\Omega + 18 \text{ k}\Omega) = 1.05 \text{ V}$$

The zener would not conduct (or fire) and V_L would be unregulated!

SAQ7.26

Yes.

$$I_L = 6.8 \text{ V}/1 \text{ k}\Omega = 6\,8 \text{ mA}$$

$$I_R = 7.33 \text{ mA}$$

(calculated in the example)

Since $I_R = I_z + I_L$, we have $I_z = I_R - I_L$. The maximum zener current I_z would now be

$$I_z = 7.33 \text{ mA} - 6.8 \text{ mA} = 0\,53 \text{ mA}$$

Therefore, $P_z > 6\,8 \text{ V} \times (0\,53\,\text{mA}) > 3\,6\,\text{mW}$.

SAQ7.27

The smoothing circuit output is 10 V with a 10% ripple. This means that $v_r = 10\%$ of $10 \text{ V} = 1 \text{ V}$ The smoothed output therefore varies from 9 V to 10 V. If the V_{in} to the regulator varies, we use the minimum V_{in} to find the value of R and the maximum V_{in} to find the value of P_z

$$R_L = 4.7 \text{ V}/0.1 \text{ A} = 47\ \Omega$$

Using the same equation as that used previously, we get $R < 43\ \Omega$. Therefore, choose $R = 39\ \Omega$. To calculate P_z, we use $V_{in} = 10 \text{ V}$ If R_L is disconnected, we have $I_z = 136 \text{ mA}$ This gives us a required power rating of $P_z > 639 \text{ mW}$.

SAQ7.28

This involves the same technique as used in SAQ7 27 This gives $R_L = 22\ \Omega$ and $R < 58\ \Omega$. If we choose $R = 56\ \Omega$, we obtain $P_z > 719 \text{ mW}$.

SAQ7.29

First we choose $R_1 = 240\ \Omega$ to satisfy the minimum I_{out} condition Next we rearrange the V_{out} equation to make R_2 the subject Therefore, we get

$$R_2 = [(V_{out}/1.25) - 1] \times R_1$$

Inserting the quantities $V_{out} = 5 \text{ V}$ and $R_1 = 240\ \Omega$, we have

$$R_2 = [(5/1.25) - 1] \times 240 = 720\ \Omega$$

We must also ensure that $V_{in} > V_{out} +$ input–output differential, i.e.

$$V_{in} > 5 + 2 > 7 \text{ V}$$

SAQ7.30

See Section 7.16 7 and the section on problems with zener voltage regulators.

SAQ7.31

Since the LM317 regulator operates by maintaining 1 25 V across R_1, we usually arrange for the current through R_1 to be slightly more than the required minimum I_{out} of 4 mA for the chip. Using $R_1 = 1.25 \text{ k}\Omega$ gives a current $I = 1.25 \text{ V}/1.25 \text{ k}\Omega = 1 \text{ mA}$ This is not enough to maintain correct operation of the circuit and is therefore not a suitable value.

SAQ7.32

Using the equation $V_{out} = 1.25(R_2/R_1 + 1)$, we have

$$V_{out} = 1.25(4800/240 + 1) = 26.25 \text{ V}$$

Since $V_{in} > V_{out} + 2$

$$V_{in} > 28.25 \text{ V}$$

Chapter 8

SAQ8.1
Since the V_{be} value should never be larger than about 0 7 V (1 V possibly for power BJTs), two of the last three BJTs must have blown.

SAQ8.2
In Example 8.2 we used worst case design to *ensure* correct operation and to fix I_b at 10 μA If $\beta = 1000$, I_b still remains at 10 μA and the BJT attempts to produce $I_c = (1000) \times (10 \mu A) = 10$ mA. However, using our R_c value at 11.8 kΩ, the BJT saturates as soon as I_c reaches 1 mA. Since saturation is a limiting condition, I_c is limited to 1 mA. All that has happened is that we have lots of base current to spare. For $I_c = 1$ mA, we actually need only $I_b = 1$ mA/1000 = 1 μA! We are therefore wasting the extra 9 μA which flows through the base-emitter junction to earth. Our circuit still works fine.

SAQ8.3
Redesign for a 20 V output merely means increasing V_{cc} from 12 V to 20 V. We need only recalculate for R_c; the rest of the working is the same as Example 8.2: $R_c = 19$ 8 kΩ and $R_b = 440$ kΩ.

SAQ8.4
Using worst case design for the BJT switch, we automatically design for a value of I_b which is probably larger than is required for correct operation. This means using a value for R_b which is probably smaller than is required. There is no problem, therefore, in choosing an R_b value *even smaller* than the circuit value. There is, however, a problem with using larger R_b values! The larger the value of R_b, the smaller will be I_b. A large R_b could easily result in insufficient I_b and insufficient I_c for saturation to occur. We therefore use the 47 kΩ resistor

SAQ8.5
The motor would take the place of R_c in the circuit. To switch the motor on properly, it requires 0.5 A. Therefore, $I_{c(sat)} = 0.5$ A. Since $I_{max} > I_{c(sat)}$, choose $I_{max} = 1$ A.

SAQ8.6
Using a lower value of R_c will probably result in the BJT not reaching saturation, since the collector voltage will be higher than before and $V_{ce} > 0.2$ V. The BJT will therefore enter its linear region and absorb more power.

SAQ8.7
Satisfy yourself that you agree with the values of R_c and R_b. To simplify the design it has been assumed that, in saturation, the voltage across the lamp is the full 24 V, neglecting $V_{ce(sat)}$.

SAQ8.8
Using the circuit of Figure 8.8, we need to determine the resistance of the lamp and its current flow We are told that $P = 22$ W and $V = 12$ V, therefore, since

$P = V \times I = V^2/R$, $R = V^2/P = 144/22 = 6$ $45\,\Omega$ This lamp will, of course, be our $R_c = 6$ $45\,\Omega$ Also, $I = P/V = 22/12 = 1.83$ A. This is our $I_{c(sat)} = 1.83$ A. Now, using $\beta_{min} = 800$, we have $I_b = 1$ $83/800 = 2.29$ mA. The base resistance is found from $R_b = (V_{cc} - 0$ $6)/I_b = 11$ 4 V/2.29 mA = 4.98 kΩ. Using any value below this gives us $R_b \leq 4.9$ kΩ.

SAQ8.9

This is almost the same method again, but this time we are not given V_{cc} and we must recognize that the LED will be in series with a current limiting resistor (see Figure 8.20). We can first assume a V_{cc} value let us choose $V_{cc} = 12$ V. We will assume also that the V_T for the LED is about 1 8 V. When the BJT enters saturation the LED lights. We need the voltage dropped across R_c. We will call it V_{Rc}. If we include $V_{ce(sat)}$, we have $V_{Rc} = V_{cc} - V_{ce(sat)} - V_T = 12 - 0.2 - 1.8 = 10$ V. R_c carries the LED current of 40 mA. Therefore, $R_c = 10$ V/40 mA = 250 Ω We are given $R_b = 40$ kΩ In this specific case the value of R_b starts out very large and reduces to a minimum value of 40 kΩ when rainwater bridges the gap between two contacts R_b is therefore shown as a variable resistance: $I_b = (12 - 0.6)$ V/40 k$\Omega = 0$ 285 mA. We can now specify our BJT β value since we know the required I_c and I_b. Therefore, $\beta = 40$ mA/0.285 mA ≈ 140. We would choose a BJT with a current gain β or $h_{FE} \geq 140$.

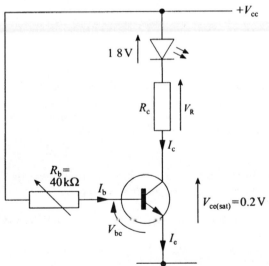

SAQ8.10

We know $I_e = I_c + I_b$, therefore, $I_c = I_e - I_b = (1000 - 10)$ μA = 990 μA and $\beta = 990$ μA/10 μA = 99

SAQ8.11

Using worst case design, we take $\beta = 100$ We are given $I_b = 0.001$ mA Therefore, $I_c = 100 \times (0$ $001) = 0$ 1 mA. $V_{Rc} = V_{cc} - 0.2 = 9$ 8 V Therefore, $R_c = V_{Rc}/I_c = 9.8$ V/0 1 mA = 98 kΩ.

SAQ8.12
Since the emitter is at 0 V, $V_b \approx 0 \ 6$ V.

SAQ8.13
The two intercepts are $V_{ce} = 10$ V and $I_c = (10 - 0.2)$ V/98 kΩ = 0.1 mA.

SAQ8.14
We would use the CC circuit.

SAQ8.15
The CC circuit.

SAQ8.16
The CE circuit.

SAQ8.17
The CE circuit.

SAQ8.18
The CC circuit has a very low output resistance and is therefore a good voltage source.

SAQ8.19
The CB circuit has a low input resistance, which means that voltage transfer to it from a high output resistance voltage source will be very poor.

SAQ8.20
Biasing is all about steady d.c values.

SAQ8.21
This is similar to Example 8.5, but this time the CC circuit requires $V_{cc}/2$ voltage drop across each of R_e and V_{ce}. Using the design steps:
1. $I_{eq} \approx I_{cq} = 1$ mA (given).
2. We have a CC circuit, therefore allow $V_{cc}/2$ (= 4.5 V) across R_e so that $R_e = V_{eq}/I_{eq} = $ 4.5 V/1 mA = 4.5 kΩ.
3 The supply voltage left is $(V_{cc} - V_{eq}) = 9 - 4.5 = 4.5$ V. This is dropped across V_{ce}. Therefore, $V_{ce} = 4.5$ V.
4. $V_{bq} = V_{eq} + V_{be} = 4.5$ V + 0.6 V = 5.1 V.
5 The lowest β is 100. Therefore, $I_{bq} = I_{cq}/\beta = 1$ mA/100 = 10 μA.
6. The voltage divider must provide 5.1 V at 10 μA.
7. R_2 carries $10I_{bq} = 100 \ \mu$A Therefore, $R_2 = V_{bq}/(10I_{bq}) = 5.1$ V/100 μA = 51 kΩ.
8. R_1 carries $11 \times I_{bq}$ and drops $(V_{cc} - V_{bq})$ volts; therefore, $R_1 = (V_{cc} - V_{bq})/(11 \times I_{bq}) = $ (9 − 5.1) V/[(11) $\times$ (10 μA)] = 35.5 kΩ.
Therefore, we have $R_1 \approx 36$ kΩ, $R_2 = 51$ kΩ and $R_e = 4.5$ kΩ.

SAQ8.22
The d.c. load line would have the intercepts $V_{ce} = 9$ V and $I_c \approx V_{cc}/R_e \approx 2$ mA.

SAQ8.23

The voltage divider is loaded by the base current into the BJT Since the base current value is dependent on the β value, the divider output will vary with β. However, since the design technique of the divider bias circuit takes into account the base current flow, variations in β will not affect the bias voltage very much. Therefore, the β dependency is minimal.

SAQ8.24

This is similar to Example 8.5 Following the same design steps will give us the following values· $R_c \approx 1\,k\Omega$ (1.125 kΩ), $R_1 \approx 5\,7\,k\Omega$, $R_2 = 1.2\,k\Omega$ and $R_e = 250\,\Omega$

SAQ8.25

The biased circuits require steady d.c. input voltages to maintain their quiescent values and to establish the Q point. Any external circuit connected to the input must not disturb the biasing conditions already set up. A capacitor does not allow a d c. current to flow through it; the capacitor will therefore block any d.c. current flow in either direction between the external input and the biased circuit This subject is covered in more detail in Chapter 9; see the section on coupling capacitors

SAQ8.26

For reasonable voltage transfer we require $R_{th} < R_L/10$. We need to find R_{th} and R_L for this circuit. The input to the BJT has a resistance value which loads the voltage divider We will call this value R_L. Therefore, $R_L = V_{bq}/I_{bq} = 1.6\,V/3\,3\,\mu A = 484.8\,k\Omega$, i.e. $R_L \approx 485\,k\Omega$. The voltage divider $R_{th} = R_1 \parallel R_2 = 369\,k\Omega \parallel 48.5\,k\Omega = 42\,87\,k\Omega$. Therefore, $R_{th} \approx 43\,k\Omega$ and $R_L/R_{th} = 11\,3$. From this we have $R_{th} = [R_L/11.3]$. Therefore, our voltage transfer condition is satisfied!

Chapter 9

SAQ9.1

We are given a Q point of (3.6 V, 1.5 mA). Therefore, we are interested only in $I_{cq} = 1.5\,mA$ and the intrinsic emitter resistance $r_e \approx 26/I_{cq} = 26/1.5 = 17.3\,\Omega$.

SAQ9.2

Given $r_e = 10\,\Omega$ and knowing that $r_e \approx 26/I_{cq}$, we can rearrange this to $I_{cq} \approx 26/r_e \approx 26/10$, giving $I_{cq} \approx 2.6\,mA$

SAQ9.3

1   $r_e \approx 26/2000 \approx 0.013\,\Omega$.

2. $r_e \approx 26/0\,05 \approx 520\,\Omega$

SAQ9.4

A mid-point biased CC circuit has $V_{ce} \approx V_{cc}/2 \approx 12/2 \approx 6\,V$. Since we are given $r_e = 52\,\Omega$, we can find $I_{cq} \approx 26/52 \approx 0.5\,mA$. Therefore, the Q point is (6 V, 0 5 mA)

SAQ9.5

The Shockley equation is $I_e \approx I_s[\exp(V_{be}/K)]$, where K is a constant approximately equal to 0.026 V at room temperature. If we differentiate this with respect to V_{be}, we will end up with an expression for dI_e/dV_{be} Since $I/V = 1/R$ (from Ohm's law), dI_e/dV_{be} gives us the reciprocal of the emitter resistance $1/r_e$. We therefore take the reciprocal of the differentiated expression to find r_e.

$$d[I_s \times \exp(V_{be}/K)]/dV_{be} = I_s \times (1/K) \times [\exp(V_{be}/K)]$$

Therefore

$$dI_e/dV_{be} = 1/r_e$$
$$= (1/K) \times [I_s \times \exp(V_{be}/K)]$$

Note that the part within square brackets is already defined as being almost equal to I_e from the original Shockley equation, giving us the simplified expression $1/r_e = (1/K) \times I_e$, where $K = 0.026$ V at room temperature, i.e

$$1/r_e = I_e/0.026$$

However, since we require r_e, we invert both sides of the expression to give $r_e \approx 0.026/I_e \approx 0\,026/I_c$ (using $I_e \approx I_c$) If we now convert to mV and mA, we end up with

$$r_e \approx 26/I_c$$

as required!

SAQ9.6

As in Figure 9.4(a), we must use $r_{in} = (\beta + 1) \times (r_e + R_e)$ and $r_e = 26/I_{cq}$ We have $\beta = 200$, $r_e = 26/10$ mA $= 2.6\,\Omega$ and $R_e = 1$ kΩ. Therefore, $r_{in} = 201 \times (1002.6\,\Omega) = 201.5$ kΩ. Using the approximation $r_{in} \approx \beta \times R_e$, we have

$$r_{in} \approx 200 \times 1\text{ k}\Omega \approx 200\text{ k}\Omega$$

SAQ9.7

We use $r_{in} = (\beta + 1) \times (r_e + R_e)$ and $r_e = 26/I_{cq}$. If $I_{cq} > 1$ mA, then $r_e < 26\,\Omega$. However, $R_e = 15$ kΩ, which renders r_e negligibly small when they are added together (r_e is less than 1% of the value of R_e). Therefore, yes we can neglect r_e

SAQ9.8

We want to increase r_e, therefore we would decrease I_{cq}.

SAQ9.9

For a grounded emitter circuit, $R_e = 0$. Therefore, $r_{in} \approx (\beta + 1) \times r_e \approx \beta \times r_e$. If I_{cq} increases, r_e decreases. In a circuit like this, a reduction in r_e causes a β times larger reduction in r_{in}. Therefore, r_{in} reduces considerably.

SAQ9.10

This question correctly suggests that r_{in} varies, as a large input signal causes large variations in collector current I_c and therefore in r_e. This can be seen from $r_{in} \approx \beta \times r_e$. The best way to stabilize r_{in} is to introduce the external R_e resistor, where $R_e \gg r_e$. The effect of variations of r_e on r_{in} is swamped by the addition of R_e in the equation $r_{in} = (\beta + 1) \times (r_e + R_e)$, where we can see that if $R_e \gg r_e$, then $r_{in} \approx (\beta + 1) \times R_e$ This is an example of adding negative feedback to the circuit.

SAQ9.11

We use $X_c = (r_{in} + r_{out})/10$ Therefore

$$X_c = 1/(2 \times \pi \times f_{min} \times C) = (r_{in} + r_{out})/10$$

Therefore

$$C = 10/[2 \times \pi \times f_{min} \times (r_{in} + r_{out})] = 10/[2 \times \pi \times 20 \times (10\,000 + 10\,000)] = 4\ \mu F.$$

SAQ9.12

$C = 10/[2 \times \pi \times 2000 \times 510\,000] = 1560$ pF.

SAQ9.13

The worst case design tells us to use $h_{FE} = 175$. Using $r_{in}' = R_1 \parallel R_2 \parallel r_{in}$, where $r_{in} \approx \beta(R_e + r_e)$ and $r_e = 26/I_{cq}$, we have

$$r_{in}' = 10\ k\Omega \parallel 10\ k\Omega \parallel \{175[1\ k\Omega + (26/4.4\ mA)]\} \approx 4\,86\ k\Omega \approx 5\ k\Omega$$

When using $r_{out} \approx R_e \parallel \{r_e + [(R_b \parallel R_s)/\beta]\}$ as we are not given R_s, we must leave it out of the expression, with the knowledge that the r_{out} value will decrease as soon as an input source is connected to the circuit. We have

$$r_{out} \approx 1\ k\Omega \parallel \{(26/4\,4\ mA) + [(10\ k\Omega \parallel 10\ k\Omega)/175]\} \approx 33\ \Omega$$

If we use the approximation $r_{out} \approx r_e$, we get

$$r_{out} \approx 6\ \Omega$$

There seems to be a large difference in the two approximate r_e values, but in reality the value of R_s is very likely to reduce the 33 Ω value to something nearer the 6 Ω estimate.

SAQ9.14

The current gain for any circuit $A_i = i_{out}/i_{in}$. In question 9.13, $r_{in} \approx 176$ kΩ and $r_{in}' \approx 5$ kΩ Thinking about this from the point of view of input current i_{in} required to drive the circuit, we could conclude that a *low* input resistance will require much more input current than a *high* input resistance. Therefore, a low input resistance will reduce the overall circuit current gain by increasing i_{in}.

SAQ9.15

Using $A_v = -R_c/(r_e + R_e)$, we have

$$A_v = -20\ k\Omega/[(26/0.5)\Omega + 2.5\ k\Omega] = -7.8$$

or more simply

$$A_v = -R_c/R_e = -20\ k\Omega/2.5\ k\Omega = -8.$$

SAQ9.16

From the expression for A_v we can see that to increase its value we need to decrease r_e. To do this we increase I_{cq}.

SAQ9.17

$A_v = -R_c/r_e = -10\,000/(26/1) = -384$.

SAQ9.18

r_e will decrease and A_v will increase.

SAQ9.19

This, in fact, is the problem with the grounded emitter circuit. Its gain varies directly with *any* variations in collector current, as you can see from the answer to SAQ9.18! To improve and stabilize this, we add our trusty external R_e resistor. Now, provided $R_e \gg r_e$, the expression $A_v = -R_c/(r_e + R_e)$ reduces to $A_v = -R_c/R_e$, which neglects r_e and its variable qualities!

SAQ9.20

You must draw an input sine wave with a 141 mV peak value and an output sine wave 180° out of phase with it, of peak value 2.83 V.

SAQ9.21

Since r_{in}' is dominated by the bias resistors, an increase in β has little effect on its value. A calculation shows that r_{in}' increases from 3 3 kΩ to 3.6 kΩ.

SAQ9.22

Since $R_c \approx r_{out}$, increasing R_c will increase r_{out}.

SAQ9.23

We need to keep R_c constant, therefore we would either increase I_{cq} (and therefore reduce r_e) or, better, simply reduce the value of R_e.

SAQ9.24

Using $r_{out} \approx R_c \parallel (1/h_{oe})$, if $h_{oe} = 5\ \mu S$, we have $1/h_{oe} = 200$ kΩ. Therefore

$$r_{out} \approx R_c \parallel (1/h_{oe}) \approx 10\ k\Omega \parallel 200\ k\Omega \approx 9.5\ k\Omega$$

If $h_{oe} = 1\ \mu S$, then $1/h_{oe} = 1$ MΩ, which has little effect on the value of r_{out}

$$r_{out} \approx 10\ k\Omega \parallel (1000\ k\Omega) \approx 9.9\ k\Omega \approx 10\ k\Omega \approx R_c.$$

SAQ9.25

In Example 8.5, values from biasing calculations were $R_e = 1$ kΩ, $R_c = 7$ kΩ and $I_{cq} = 1$ mA. We therefore split R_e into R_{e1} and R_{e2}. R_{e2} will be the lower resistor with a bypass capacitor connected across it. Since we arrange for R_{e2} to be bypassed by the action of C, this means that R_{e2} is effectively removed from the gain equation, leaving us with

$$A_v \approx -R_c/(R_{e1} + r_e) \approx -20$$

We can find r_e from $r_e = 26/1 = 26\ \Omega$. Rearranging for R_{e1}, we have

$$R_{e1} \approx -[(R_c/A_v) - r_e] \approx 324\ \Omega$$

(The minus sign at the front of the equation is no use here.) We also know that

$(R_{e1} + R_{e2}) = 1 \, k\Omega$. Therefore

$$R_{e2} = 1000 \, \Omega - 324 \, \Omega = 676 \, \Omega.$$

SAQ9.26
Using $X_c = R/10$, where $R = 900 \, \Omega$ and $f_{min} = 2 \, kHz$, we have

$$1/(2 \times \pi \times f_{min} \times C) = R/10$$

Therefore

$$C = 10/(2 \times \pi \times f_{min} \times R) = 884 \, nF.$$

Chapter 10

SAQ10.1
The E-MOSFET in this case requires a V_{gs} of +2 V or more for inversion to take place. Therefore, the channel will not be inverted

SAQ10.2
We need to use the expression

$$I_d = I_{d(on)} \times [(V_{gs} - V_{TH})^2/(V_{gs(on)} - V_{TH})^2]$$

For $V_{gs} = +8 \, V$, where $V_{TH} = +2 \, V$, $I_{d(on)} = 1 \, mA$ and $V_{gs(on)} = 5 \, V$, we will have

$$I_d = 1 \, mA \times [8 \, V - 2 \, V)^2/(5 \, V - 2 \, V)^2] \approx 4 \, mA.$$

SAQ10.3
The gate current to any MOSFET is zero!

SAQ10.4
We need to check which of the following expressions is true:
- If $V_{ds} < (V_{gs} - V_{TH})$ it is in linear or ohmic mode
- If $V_{ds} > (V_{gs} - V_{TH})$ it is in saturation or current mode.

We are given $V_{ds} = +3 \, V$, $V_{gs} = +4 \, V$ and $V_{TH} = +2 \, V$. Therefore, $(V_{gs} - V_{TH}) = (4 - 2) = +2 \, V$. Since $V_{ds} > +2 \, V$, the MOSFET is clearly in the saturation (or current) mode

SAQ10.5
To remain in ohmic mode, $V_{ds} < (V_{gs} - V_{TH})$. We are given $V_{ds} = +10 \, V$ and $V_{TH} = +3 \, V$. Therefore, $10 < (V_{gs} - 3)$, which gives us $(10 + 3) < V_{gs}$ or $+13 < V_{gs}$. This means that V_{gs} must always be more than $+13 \, V$.

SAQ10.6
To calculate any R value we need V and I, and in this case, V_{ds} and I_d. From Ohm's law, $R = V_{ds}/I_d$ In the saturation region, I_d stays constant Therefore, as V_{ds} increases, so does R.

SAQ10.7
For any plot of I against V (where I is the vertical axis), the steepness of the

gradient will indicate the resistance value, i.e. the steeper the gradient, the lower the resistance (see Section 7.12 on diode resistance for an explanation). In Figure 10.6(b) we can see that as V_{ds} increases, steepness increases slightly and therefore the resistance decreases slightly.

SAQ10.8
$I_d = I_{dss} \times [1 - (V_{gs}/V_{gs(off)})]^2 = 5\,\text{mA} \times [1 - (+6/-4)]^2 \approx 31.25\,\text{mA}$.

SAQ10.9
Zero.

SAQ10.10
If V_{gs} was reduced to $V_{gs(off)} = -4\,\text{V}$ or lower, the device would be switched off

SAQ10.11
Because $V_{gs} = -3\,\text{V}$ is more negative than $V_{gs} = 0$, we have $I_d < I_{dss}$.

SAQ10.12
An enhancement device.

SAQ10.13
Enhancement mode.

SAQ10.14
Referring to the NMOS circuits in Figure 10.12, I_d flows whenever $V_{out} \approx 0$

SAQ10.15
I_d *only* flows when the two devices are switching either on–off or off–on. This happens very briefly and I_d drops to zero immediately afterwards. For most of the time, $I_d = 0$.

SAQ10.16
Yes, as illustrated in the case of the CMOS inverter circuit in Figure 10.13.

SAQ10.17
This is basically bookwork, but it is useful knowledge to have at your fingertips.
1   A   MOSFET.
2   A   BJT.
3   A   BJT.
4. A   MOSFET.
5   A   BJT.
6. A   BJT.
7   A   BJT.
8. A   MOSFET.

SAQ10.18
Whenever a resistance is needed in an integrated circuit, much less silicon area is

taken up by an active device (e g. a MOSFET or a BJT) than a pure resistor Therefore, active devices (connected as resistors) would always be used if possible.

SAQ10.19
A p channel E-MOSFET requires a $V_{gs} \leq V_T$ to switch it on This device has $V_T = -3$ V, therefore $V_{gs} = 0$ is insufficient and the MOSFET is off.

SAQ10.20
An op-amp is required to have very high R_{in} and A_v. MOSFETs would give a high R_{in}, followed by BJTs which give a high A_v.

Index